COLUMNS

COLUMNS

JONATHAN ADAMS

Detail in Building

A.D. ACADEMY EDITIONS

Detail in Building

Advisory Panel: Maritz Vandenberg, Christopher Dean, Christopher McCarthy, Maggie Toy
Consultant Editor: Michael Spens

Dedicated to Hazel and Harri Taliesin Adams

ACKNOWLEDGEMENTS
The author would like to thank all the architectural and engineering practices that have contributed material to this book, in addition to the following individuals: Benedict O'Looney, Roderick Coyne, Tony Reason, Martin Wood, Aker Maritime, the Commissioners for Irish Lights, the Wind Energy Group, and SJ and Dr MS Adams.

Picture Credits
Every attempt has been made to locate the sources of all images used in this publication to obtain full reproduction rights but in the very few cases where this process has failed to find the copyright holder, apologies are offered. For the purposes of these credits, images in the first part of the book are referred to by figure numbers alone; those in the case studies by folios and figure numbers. Jonathan Adams (figs 4, 5, 8, 9, 11, 12, 14, 15, 23, 45, 46); Jonathan Adams and B O'Looney (page 62, figure 2; page 63, fig 6); Almasy (fig 6); Tadao Ando (fig 42); Claudia Andujar (fig 34); James Ardiles-Arce (fig 47); from *Architectural Design: Britain in the Thirties* (fig 98); Tim Benton (fig 54); Noel Le Boyer (fig 16); Bibliothèque de la Ville La Chaux-de-Fonds (fig 51); Ricardo Bofill (figs 66, 67); Richard Bryant (fig 72); Pierre Joly and Vera Cardot (fig 48); Peter Cook (pages 92–5, figs 1, 4, 5); Peter Durant (page 91, fig 4); Editions du Pont Royal (figs 10, 18); Arne Peter Eggen (fig 44); Gisèle Freund (fig 76); Sigfried Giedion (from *Space, Time and Architecture*, Harvard University Press, 1962) (fig 31); Institute of Civil Engineers (fig 86; page 62, fig 4); Ironbridge Gorge Museum Trust (fig 26); H Lacheroy (fig 87); Carl Laubin (fig 74); Leicester City Museums (page 60); Isabelle Lonsada (fig 89); The Louvre, Paris (fig 2); André Jammes (fig 27); Johnson Wax Corp (fig 43); Johnson Wax Corp/Susan Wolderberg (fig 39); Hasan Udin-Khan (fig 18); *New Welsh Review* (fig 78); Marburg Institute (fig 8); Max Fordham Engineers (page 92, figs 2, 3) Victoria Melnecke-Berg (fig 96); NASA (fig 99); Lennart Olsen (fig 52); M Ottin (fig 13); Renzo Piano/Richard Rogers (fig 61); Stephen Pimbley (page 75, fig 8); P Popper (fig 77); Sunand Prasad (fig 50); Paul Raftery (front cover image); Francois Réné Roland (figs 36, 37, 38); Rome Tourist Office (figs 80, 81, 82); John Romer (fig 91); Joseph Rykwert (from *The Dancing Column*, MIT, 1996) (figs 19, 20, 21, 33); Alain Salaun (fig 58); Jorg Schlaich and Gunter Mayer (fig 97); Curt Seigel (fig 53); Sir Norman Foster and Partners (figs 63, 64); James Stirling (fig 49); Spanish National Tourist Office (fig 15); Otto Steidle (figs 69, 70); Giulio Veggi (figs 7, 84); Robert Venturi (fig 71); Villa Mairea Foundation/ Rauno Traskelin (figs 55, 56, 57); B Vincent (figs 61, 62); Deidi von Schawen (fig 68); Morley von Sternberg (page 88, fig 1; page 90, fig 7); John Edwin Wood (from *Sun, Moon and Standing Stones*, OUP, 1978) (figs 3, 75); Michael Waddington (page 80, fig 4).

Cover: Alsop & Störmer, Hôtel du Département des Bouches-du-Rhône, Marseilles, 1990–95 (photo: Paul Raftery)
Page 2: Herron Associates, Kurobe Wind Tower, 1992

First published in Great Britain in 1998 by
ACADEMY EDITIONS

A division of
JOHN WILEY & SONS
Baffins Lane
Chichester
West Sussex PO19 1UD

ISBN 0-471-97807-8

Other Wiley Editorial Offices
New York • Weinheim • Brisbane • Singapore • Toronto

Printed and bound in Italy

CONTENTS

FOREWORD

As the Detail in Building series has developed it has become clear that each subject covered requires a slightly different treatment. Nowhere is this more apparent than in the case of the column which has long been considered more than just a structural detail, forming the basis of Western architectural systems and theories since the Renaissance. In his work for Alsop & Störmer, Jonathan Adams has had countless opportunities to explore the column's uses and possibilities, and his broad range of experience informs this expansive and lively treatment of the subject. What has emerged is a slightly more text-oriented presentation than previous titles in the series, and the design of the book has been adapted accordingly.

The first section of the book consists of four chapters which present the history of the column – from prehistory to the close of the twentieth century – and the ideas and theories that it has engendered, including detailed studies of a fascinating range of key examples. The second section of the book presents ten case studies which demonstrate the scope and innovation of this element in building today.

Adams conveys the tremendous breadth of the subject, investigating not only structural or supporting columns, but also monolithic or freestanding columns. In the case studies as well as the main text, he demonstrates the diversity of the column. Examples include the Kish Bank Lighthouse, the Troll Gravity Base Structure in the North Sea, the MS3 Wind Turbines, and the Surge Column which also functions as a barometer for London's commuters. *Maggie Toy*

INTRODUCTION

It would be absurd for any great city to be without an obelisk. Rome has had them this great while and so has Constantinople. Paris has one. London has one. If New York was without one [they] would point the finger of scorn at us and intimate that we could never rise to any real moral grandeur until we had our obelisk.

New York Herald, January 1881

1 Leonardo da Vinci, *John the Baptist* (detail), 1515 (courtesy of The Louvre, Paris)

The American Cleopatra's Needle arrived in Manhattan within weeks of the *Herald*'s popular proclamation. Five years after the 'moral grandeur' of New York had been secured, Washington went some way further by putting the finishing blocks on the point of the Capitol Monument, 170 metres up, higher than any other freestanding column.

Moral grandeur is a difficult virtue to define, let alone to express; but of any lasting gesture that has evoked such a lofty notion none is so intuitive, so exhilarating or has endured as long as the monumental column. Since the prehistoric era these strange, purposeless gestures have been synonymous with civilised aspirations, and it is certainly true that the column's persistent association with the higher ideals of humanity far outweighs the significance of any constructional detail. Although they may have come superficially to mean different things, the freestanding columns and the structural props that proliferate in our buildings are fundamentally the same; they share the same origin and are invested with the same primal significance.

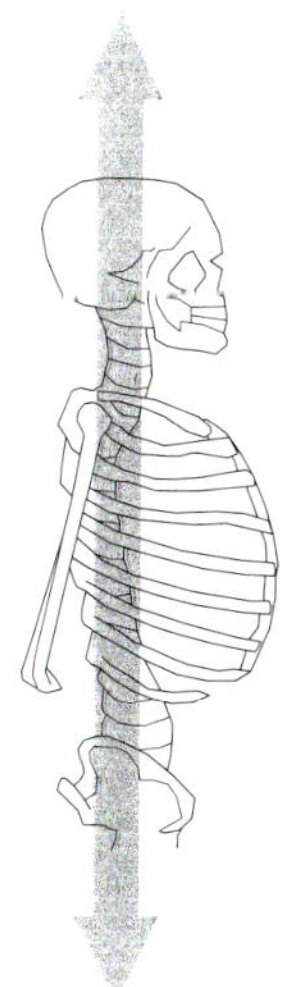

2 The human spinal column: upright posture

It may be difficult for us to imagine architecture without columns, simply because we recognise their utility above their symbolic value. Architects use columns to open up floor space and to elevate horizontal planes. We are used to thinking of them as the logical means of achieving such spatial effects. Of course it does no harm to take them for granted in this way, but if we do, then perhaps we are contributing to the empty pragmatism of modern building – the tendency that reduces our art to a business alone and our buildings to no more than capital assets.

Columns provide a generally untapped source of mythic energy, a source that is available to any designer of buildings. They represent the vertical features that once occurred only in nature, the tall rocks and straight trees that were venerated

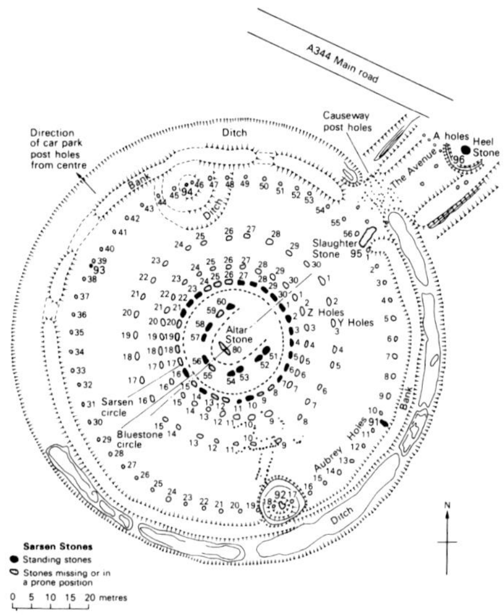

3 Plan of Stonehenge, Wiltshire, *c*2600–1800BC

by our forebears simply because they drew a line through the plane of their day-to-day existence. In many cultures, with varying degrees of abstraction, columns came to represent the human figure or the individual human being. They echoed the divine connotations of the index finger pointing upwards and of the spinal column and its upright posture invested uniquely in mankind by his Creator. Whatever the route from living tree to man-made monolith it is the simple verticality of the column that is most compelling as it is this that most clearly embodies the primordial yearning of our forebears to be closer to the heavens.

We seem, also, to have been equipped by evolution to discern geometrical relationships from background chaos. Our prehistoric ancestors used patterns to define shared frames of reference which would in turn instil a sense of order and common purpose among their communities. For thousands of years before they were used as supports columns were commonly erected in alignments, grids and encirclements which related to perceived calendrical and other supernatural orders. The narrative progression of placements and the rhythmic spacing of verticals are idealised expressions of longed for natural order; they are the antidote to all that is wild, dangerous and unpredictable in our world.

The Italian Renaissance of the fifteenth and sixteenth centuries was characterised by a revival of interest in the cultures of Hellenic Greece and Rome. Building had flourished throughout Europe since the decline of the Roman Empire, but it was not until the fifteenth century that it occurred to Italian scholars that the building of the ancients represented an aesthetic purity that had been lost through the Dark Ages. The revival of the 'classical' manner introduced the notion of the 'Architect' – the master builder. The sole qualification of the true master builder was that he should possess the formulae of the three basic column types that became known as the Greek 'classical orders', and that he should be able to execute them faithfully in accordance with ancient patterns. The architect was concerned with scholarship more than he was with technique, and the practice of architecture was to bring the mastery of building out of the mason's yard and into the library and studio. The association of building design with theory and esoterica is responsible for the distance between architects and the building industry that persists to this day.

It is only proper that this essay should begin with a more detailed analysis of the origins of the classical columns and the legacy of the beautiful delusion that they inspired.

Architects today clearly regard it as helpful to include all and any built structures into the gamut of their responsibilities, and specific classifications of contemporary column 'orders' are likely to be as useless as they are anachronistic. Contemporary design is increasingly driven by technology. Designers frequently make a virtue of the execution of details at the expense of an holistic consideration of form and space.

Furthermore, it is common to hear architects speak of their work in terms of 'problems' and 'solutions'. Such supposedly 'objective' reasoning is used to suggest that ideas rendered in architecture are somehow the product of circumstance and that they cannot really reflect our imagination. If there is something to be gained from Renaissance architectural thinking it is not that we should yet again revive the classical style but that we should embrace the fact that if our design speaks for us it can speak of a higher purpose than mere utility. Later in this essay there is a survey of most of the more-or-less contemporary column forms, but these are presented with the intention of encouraging the reader to add to the canon rather than to draw from it.

Columns are amongst our most potent expressive devices. The detail of their manufacture may be of some interest when scrutiny of the minutiae contributes to an understanding of the whole, but it shouldn't distract us from the more expansive levels of subtlety which are accessible to us through the thoughtful design and placement of the column. The detail of columns is usually quite obvious – it is the *gestalt* which holds the mystery.

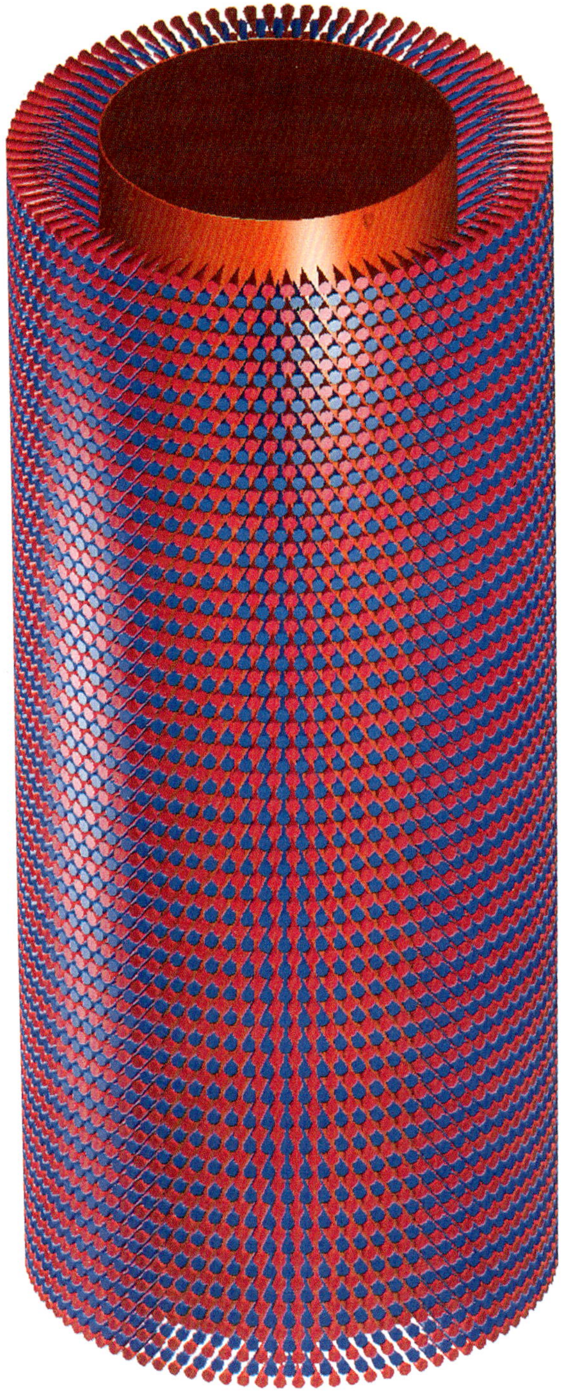

4 Computer model of Uruk cone mosaic column, *c*3000BC

CHAPTER 1

THE CLASSICAL COLUMN AND THE ORIGIN OF ARCHITECTURE

BEFORE VITRUVIUS

The sacred grove was by no means a substitute for the temple: the wood was *the temple, its trees the columns and the firmament its roof. The word 'templum' signifies a . . . piece of land dedicated to godhead, a holy precinct. Most houses of god betray their vegetable origins by being oriented, and opening up to, the sun. Thus what we call a temple is actually an abstraction of a grove, the thicket of columns recalls the thicket of trees.*

Bernard Rudofsky, *The Prodigious Builders* (1977)

5 Colonnade of treetrunks

Architecture originates with the notion that aesthetic perfection was achieved in the Hellenic period (*c*323–27BC), and that all subsequent practice has been an attempt to retrieve that state of grace. For this reason, Hellenic building tends to be thought of as a beginning, rather than as a relic of a transitory culture. However, in the millennia before the Greek and Roman empires, there was far more diversity of design than the limited remnants of those cultures suggest.

It seems obvious that if all building forms and components share a 'common ancestor' then this is the tree trunk (5). Quite apart from the heady spiritual potency of the tree, from ancient times until the present, timber has been the most useful and sympathetic building material available to us.

Trees are friable. Once cut and erected they are prone to decay rapidly. Stone columns, however, will, like the living tree, outlive any man. A thousand years before the Hellenic period, the columns of the Minoan and Mycenaean civilisations were each cut from individual tree trunks which, probably to prevent them taking root, were erected head down so that they tapered outwards towards the top. Furthermore, they were thickly painted for decorative and preservative purposes, characteristically with a red shaft and black capital. The remains of the vast palace complex at Knossos in Crete (*c*2150/2000–*c*1300BC) attest to a society equally skilled at working stone, metals and timber on the grandest of scales (6). The use of timber for the columns endured not because of a lack of alternatives, but because wood was evidently believed to be the *appropriate* material for the purpose.

6 Columns at Knossos, Crete, *c*2150/2000–*c*1300BC

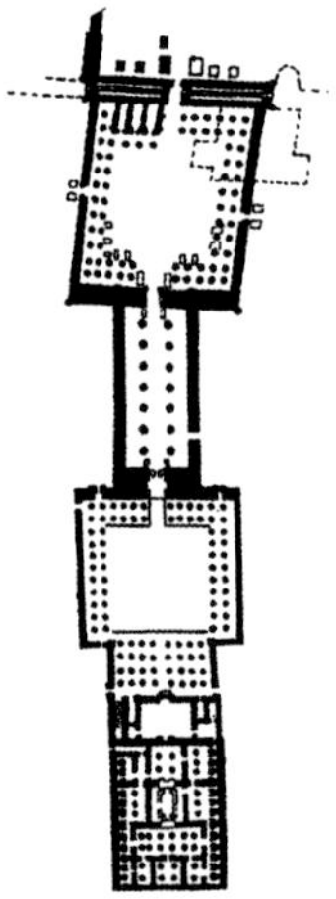

7 Temple at Luxor, Thebes, 1408–1300BC

8 Great Temple of Amum, Karnak, Thebes, 1530–323BC

The splicing and wedging techniques used for jointing the beam timbers into those of the roof probably continued unchanged for centuries. Used in the earliest Hellenic temples, which are known to have been fully and partially timbered at various times, they survived as the curious triglyph entablature ornament seen around the roofs of Greek temples. In the same way, it requires no great imagination to deduce that the Doric capital represents some kind of capping vessel or binding that had once been used to seal or ornament the head of the cut timber shaft. For the Greek builders, these vestiges of a sacred building tradition were significant enough to merit retention long after the appropriateness of real timber became a cultural anachronism.

In their technical ambition and expressive vitality, the hypostyle columns that survive from the Egyptian Middle Kingdom (*c*2008–*c*1630BC) are surely the equal of any other relics from prehistory or since (7, 8). The tree representations are much more overt and unencumbered by evolved symbolism than the later Greek forms. Despite always being arranged on tight orthogonal grids, and always sharing the same swollen profile, the columns support a great variety of representational capitals illustrating several distinct tree species. Egyptian society, perhaps to a greater extent than any other, organised itself around the execution of ambitious masonry engineering projects. Aside from the familiar giant arboreal forms, there are examples of apparently proto-Doric columns from much earlier dynasties: square columns, figurative columns, free-standing obelisks and a vast diversity of other expressive traditions in the medium of stone.

It is tempting to assume that all prehistoric civilisations began their serious building in timber before graduating to worked stone, and that load-bearing masonry in columns is a specification that has never been bettered. However, the sometimes suspiciously neat assumptions of 'classical history' do not always account for the evidence. Remains from the ancient Uruk civilisation (a forerunner of later Mesopotamian culture) in the Eanna temple complex, Sumeria, show that here load-bearing columns were not of simple timber construction, but were of what we might now consider a highly 'contemporary' composite of stone, cement and ceramic mosaic. Today, glass or ceramic mosaic tiles are commonly used as a skin over the brawn of reinforced concrete. In the Uruk columns, the visible mosaic disc was not a superficial tile but the circular end of a coloured ceramic cone. The points of the cones were embedded in a closely packed hexagonal pattern 5–10 centimetres into the cementitious casing of the rough stone column, giving a secure patterned finish which was also a reinforcing aggregate for the cement matrix (4). This technology was developed around 3000BC, over a thousand years before Stonehenge was built.

THE ACCIDENTAL LEGACY OF VITRUVIUS

The origin of the classical orders is lost to us, if it was ever known. The three Greek orders – the Doric, the Ionic and the Corinthian – developed over millennia and across thousands of miles of the prehistoric Mediterranean. It was not until *c*40BC that the Roman Marcus Vitruvius Pollio described the forms of Hellenic building in his treatise *De architectura* with the intention of bestowing some of the legendary moral loftiness of the culture of Alexander the Great upon an increasingly dissolute Roman society. As it was written some time after the height of the Hellenistic period, we may assume that much of Vitruvius' lyrical documentary is fable or folklore.

The first translation of Vitruvius into Italian was printed in 1521, less than a century after the book's Renaissance rediscovery. By this time, acceptance of its veracity had already firmly established architecture as one of the highest expressive art forms, and its exclusive rules, based on the classical orders, were literally cast in stone.

Although the orders appear superficially to amount to little more than decorative patterns, they were, according to Vitruvius, permeated with a system of geometries that expressed the relationship between the human body and the underlying principles of mathematics. This would, to Vitruvius, have been part of a familiar and ancient rubric of human proportion. To the monotheistic scholars of the Renaissance, it was a resounding affirmation of contemporary philosophical and theological speculation which linked human proportions to certain geometric absolutes. Renaissance art and literature were obsessed with the recognition of divinity in the everyday and most emphatically in the human form, in its symmetries and proportions. Leonardo da Vinci's famous *L'Uomo Universalis* sums up the efforts of many of his contemporaries to portray the human body as a microcosm of the universe, capable of infinite subdivision into mathematically significant portions, rather as a Mandelbrot set reveals progressively smaller images of itself (9).

Whatever his original intentions, Vitruvius' treatise raised, for his adherents, the previously undreamed of possibility that building could literally embody divinity, not by the beseeching structural gymnastics of the gothic idiom, but by the simple means of a correct organisation of columns and voids.

Renaissance classicism is not so much a revival of Greek manners as it is a revival of the Roman interpretation of the Greek way. In Vitruvius, the three true Greek orders are supplemented by a Roman variant: the Tuscan. The inclusion of this fourth form introduces to the classical canon the characteristically Roman regard for status. Whereas in Greek building the Doric is without question the most noble of forms despite its relative austerity, in Roman classicism, it is the very plainness of the Tuscan that makes it specifically suited to mundane applications – basements, servants' quarters and so on – and the Doric rates just above it in status.

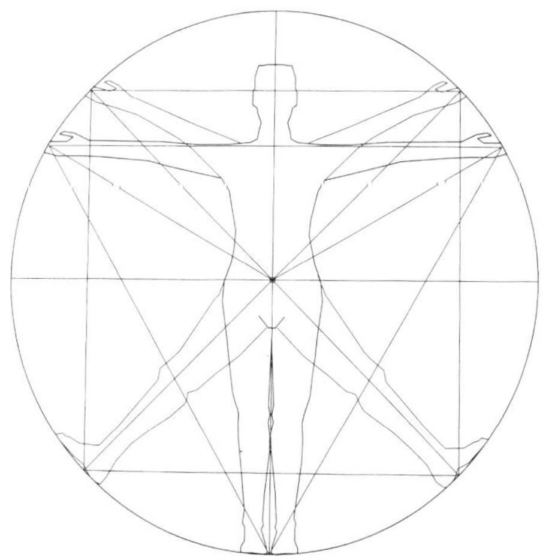

9 *L'Uomo Universalis*, drawing after Leonardo da Vinci

10 Doric columns at Paestum, Sicily, 6th century BC

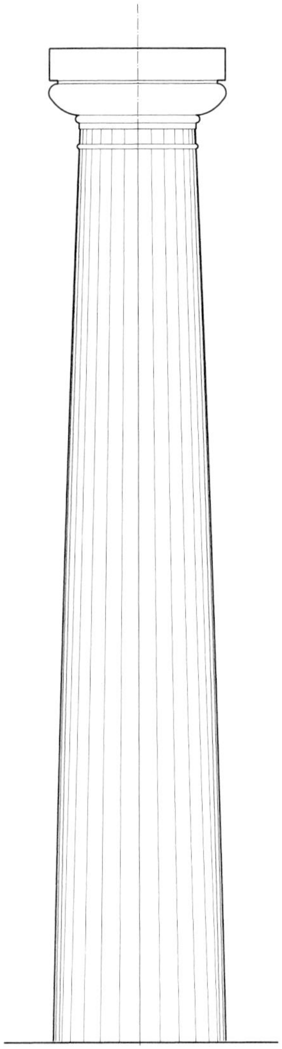

11 Doric order

Vitruvius also indicated the appropriate applications for the three true orders and his Roman grading procedure was picked up by Renaissance theorists who had the problem of deciding which column forms to apply to each of the multitude of their new building types.

A fifth order, the Composite, was added to the formal classical canon by Sebastiano Serlio in the sixteenth century. A Roman hybrid of the Ionic and Corinthian, although it was popular with the builders of the grand Roman projects, it could not be said to be a distinct form and was not defined as such by Vitruvius. The literature of classicism is vast: arguments over the relative merits of the orders and other trifling areas of interpretation have preoccupied architects for centuries. From our perspective, the Roman hybrids are significant mainly for establishing the specious but persistent idea that the orders lend themselves to regionalisation – an 'American order', a 'Japanese order', a 'New Delhi order', etc – and that this is a worthwhile exercise in moral aggrandisement for those intent on demonstrating that they and their society are the equal of any that preceded them.

The Doric Order

Not until 1996 has a close-to-definitive treatise on the origin of the classical orders been written: this is Joseph Rykwert's *The Dancing Column*. In this book, Rykwert describes the Doric as 'a grandiose and primal form that seems to spring directly from the instinct to build with such vigour that it had to transcend necessity'; it is the oldest of the orders, and also the most abstract and refined (10, 11).

The fact that it seems more remote from its 'vegetable origins' than the more tree-like Ionic and Corinthian forms suggests that something other than the obvious tree references must account for the two latter forms, and that, between Rudofsky's poetic forest glade and the stone temples that remain as relics, there was an indeterminate phase of temple building in timber. During this period, the temple moved from the wilderness into the settlement, and the living trees were cut and worked into timber columns. The Doric is the stone model of those timber columns. The features of the column and the entablature above represent the simple components of a wooden building rather than a forest glade.

According to Vitruvius, the first Doric temple was built by emigrants from Athens, in order to lay claim to newly colonised lands. The temple was built, by popular consensus, to resemble a style that the travelling community had seen used in the Dorian-speaking islands of south-western Greece, and it was dedicated to Apollo. Without a model to hand, the builders settled on the proportions of the ideal male body, as appropriate for Apollo. This proportion was the ratio of the length of the foot to the length of the body, in the masculine case 1:6, and the shaft was boldly fluted.

The Ionic Order

As a neat parallel to the Doric myth, according to the Roman storyteller, the first Ionic temple was dedicated to Diana, and embodied female proportions: 1:8. The Ionic column shaft is always delicately fluted, perhaps to represent the pleats or folds of the light garments usually worn by the female Grecian deities (12, 13, 23). From these observations stems the notion that classical columns are perhaps primarily figurative. Furthermore, whereas the Doric might have been held to represent human proportions generally, by ascribing specifically feminine properties to the Ionic, the masculinity of the Doric was emphasised. Hence the orders can be considered in terms of gender.

Vitruvius speculated that the distinctive volutes – the two mirror-image downward-spiral shapes that adorn the capital – represent the curls of the huntress' hair. The volutes do seem to present one of the most tantalising puzzles of the classical orders: they so obviously represent something, though exactly what they represent has never been satisfactorily identified. Rykwert's survey of Ionic variants and proto-Ionics suggests a many-layered metaphor: the horns of sacrificial beasts, the boughs of sacred trees and, indeed, the alluring curls of the female goddess or priestess.

Whatever the specific origins, the femininity of the Ionic, as inferred by the masculinity of the Doric, seems to have been accepted by Vitruvius and by every scholar to have interpreted his work.

12 Ionic order

13 Ionic columns, Temple of Apollo, Yenihisar, Turkey, 4th century BC

The Corinthian Order

It seems paradoxical that the most obviously representational of the three orders, the Corinthian, is generally considered the least edifying. It is thought to be the youngest of the orders and, for that reason perhaps, the least refined (14).

The mop of foliage that comprises the bulk of the capital is known to represent the acanthus plant. The acanthus has some sacred significance; many objects were so endowed in ancient Greece where multitudes of religious cults practised side by side. The acanthus appears to grow out of a bowl, and from the top of the bush, vestigial volutes curve outwards on the diagonal axes. Vitruvius' account of the mythical origin of the Corinthian form is comically implausible (an accidentally upturned pot found by a roadside is a pivotal incident): Rykwert's book relates this and other myths in mesmerising detail.

Of relevance in its relationship to the other orders, however, is Vitruvius' observation that the Corinthian proportions (1:10) are also derived from the female body – in this case the more slender body of an adolescent maiden, the chosen form of the goddess Aphrodite. Male/female has a pleasing symmetry which is disturbed by the addition of a second female. It is probably the gender imbalance that has tended to devalue the Corinthian relative to the two former types.

14 Corinthian order

AFTER VITRUVIUS

15 Medieval grain store, Galicia, Spain, 12th century

16 Nave of Coutances Cathedral, France, *c*1220–91

17 Hall of Columns, Great Mosque, Cordoba (961–65)

18 Lion Court, The Alhambra, Granada (principally 1338–90)

The history of building in Europe between the end of the Roman Empire and the dawn of the Renaissance is not really an architectural history. Indigenous vernaculars evolved very slowly around an unchanging palette of unrefined and generally perishable materials. The big projects were castles and fortifications, the property of the gentry, and ecclesiastical buildings for the Church and its many offshoots (15).

The workers of stone, the masons, organised themselves into guilds which wielded considerable influence over the economics of medieval Europe. It is through their movement across the Continent that one can follow the evolution of building techniques which culminates, in the view of most historians, in the high gothic cathedrals of northern Europe (16). Boundaries between Christian and Islamic territories were flexible throughout this period; places of worship often changed hands and, as a result, there is much greater commonality between the medieval building forms of the two cultures than is apparent today (17, 18).

Although a few manuscripts of Vitruvius were known and used, only traces of the influence of the classical orders can be discerned in medieval building. What does survive, perhaps because it seems natural enough to be made a convention regardless of its origins, is the habit of providing the vertical support with a head and a base.

In Christian buildings, the head of the column is typically adorned with representations of animals, plants and saints. Solid stone was still the material of choice. Without theory or calculation, and despite occasional catastrophic collapse, the structural gymnastics performed by the gothic masons, using generally unfixed stone within the confines of simple compression systems, defy belief to this day.

Unadorned stone was also the preferred medium of the first classical revival. The directions taken by the newborn art of architecture were determined by the dissemination of Vitruvius and, in the sixteenth and seventeenth centuries, a burgeoning scholastic industry of Vitruvian interpretation. Within a few generations these manifestos had resulted in the abandonment of the structural daredevilry of the gothic style and the eclipse of the masonic hegemony by the exclusive clique of 'higher' architects.

The earliest book on architecture printed in English was John Shute's *First and Chief Groundes of Architecture* of 1563. This book is almost entirely concerned with illustrating the Vitruvian mythology about the parallel between the orders and their human body equivalents (19, 20, 21). The importance of these analogies cannot be understated because it is this relationship which, as has been explained, conveys the influence of the divine hand in the creation of the orders. The same theme was central to Roland Fréart de Chambray's influential textbook *Parallele de l'architecture antique et de la moderne* of 1650.

The most influential text of this period was Andrea Palladio's *Quattro Libri dell'Architettura,* published in 1570. Intended partly as a catalogue of Palladio's many, justly celebrated, works, it also contains an exposition of the orders which was practical rather than scholarly. While providing helpful technical hints, such as how to bend a lath to make a template for the swelling (entasis) of a column, Palladio was emphatic that the forms of the orders were not open to re-interpretation, and that any deviation from the pattern was presumptuous, vulgar and certainly not 'architecture'. Curiously, while he would not tolerate it in the work of others, Palladio frequently bent the rules himself and his work was, at various times in later centuries, severely criticised for its 'mannerism' by more hard-line classicists. As we will see, this dichotomy recurs throughout architectural history: the visionary architect harks back to the classical 'origin' of building, elevates it in theory to the highest status, claims it on behalf of a new generation of architects for a new society, and then happily does his own thing (22).

Inigo Jones (1573–1652) is credited with having introduced the *Quattro Libri* to England in the seventeenth century. On the Continent, Palladio's treatise had for some decades been the standard text, backed up by his impressive body of building. It was not until the 1740s, however, that demand for an English translation was eventually met, and by that time his inspired Vitruvian poeticism had already been universally adopted throughout Britain and America.

Christopher Wren (1632–1723) was, in his own lifetime, considered the greatest architect of the English-speaking world. Whilst always characteristically his own, Wren's architecture was not however technically daring or innovative. It was a return to aesthetic perfection rather than innovation that was sought through mastery of classical language. As late as the mid-eighteenth century, John Wood the Elder, the great architect of Georgian Bath, stated his conviction that classicism owed its origins to the divine revelation of the three orders to the Jews for the Temple of Solomon (*The Origin of Building or the Plagiarism of Heathens Detected*, 1741).

If there is little variety in the techniques of column construction between the Renaissance and the nineteenth century, it is because the principles of architecture demanded that it should be so.

19 The origin of the Doric order, from John Shute's *First and Chief Groundes of Architecture* (1563)

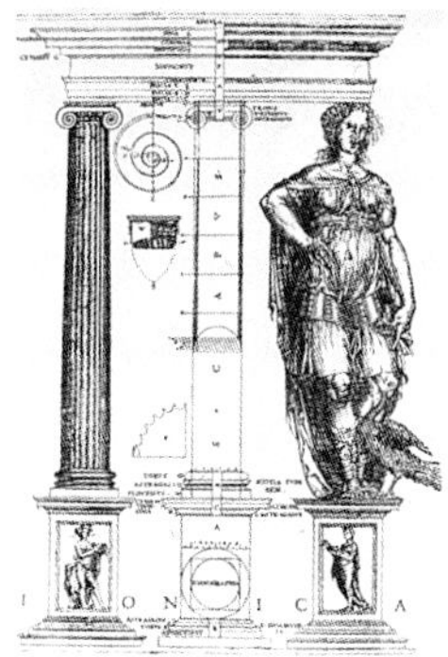

20 The origin of the Ionic order, from John Shute's *First and Chief Groundes of Architecture* (1563)

21 The origin of the Corinthian order, from John Shute's *First and Chief Groundes of Architecture* (1563)

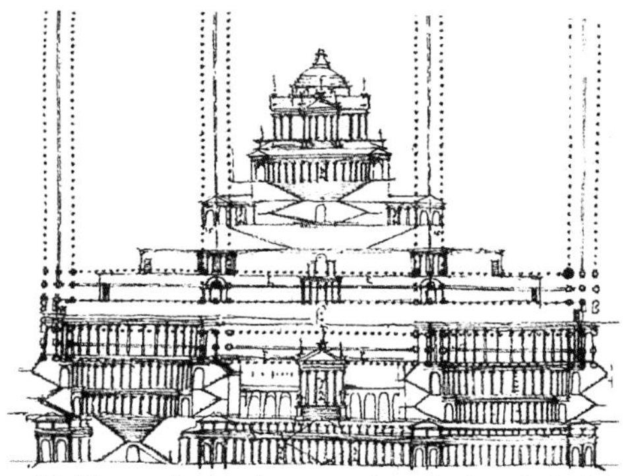

22 'Roman Fantasy', a page from Andrea Palladio's sketchbooks, *c*1550

VITRUVIUS NOW

Revival of interest in the classical orders has recurred cyclically since the Renaissance. Typically, there will be an inclination to return to 'origins' whenever a crisis of confidence afflicts society.

Today we work within a culture of building dominated by exact repetition of components and in which 'economy of means' is considered an aesthetic.

Architectural theorists at the beginning of this century were enraptured by technological advance and readily subscribed to the idea that 'superfluous' design was ornamental, economically and technologically unsound, ill-suited to mass production and therefore aesthetically inappropriate. Despite its claim to be a 'classical' form, the legacy of the Bauhaus is an architecture which is isolated from its community, 'crystalline' and invulnerable. For the type of designer who feels exposed without a prescriptive and easily attainable absolute to which to aspire (and for whom the respect of fellow ascetic missionaries usually far outweighs the value of popular understanding), the minimalist vision is one that still holds a powerful allure, and the architecture of ancient Greece is, for some reason, still regarded by minimalist designers as the paradigm: we take from classicism that which we wish to find.

For its part, with or without the encouragement of statesmen and royalty, the public is sympathetic to the proposition that the sameness and regularity – the purity – of Greek architecture is in accord with both modern conservatism and moral enlightenment; that unadulterated classicism offers, in effect, a panacea for the crisis of urban building.

The evidence suggests, however, that the Greek builders would not have recognised an association between their work and twentieth-century minimal functionalism, or for that matter with twentieth-century neo-classicism.

Over the bones of the classical orders the Greeks plastered layers of gaudy stucco and paint. Rather than the dumb, priapic colonnades beloved of proto-fascists, the Greek builders organised the geometries of their buildings to capture sublime abstract forms: they curve and converge in ways which might be regarded as daringly contemporary, parabolic curves so shallow that they flirt with invisibility, columns that lean progressively in towards the centre line of the building, each one a degree or two more than the next. Furthermore, there is little exact similarity between the columns of different buildings, even between examples of the basic Doric order. The literature insists that there are three types only: it appears that there are certainly three generic forms, but there exist within the Hellenic canon plentiful hybrids and proto-forms to blur the boundaries between them. Neither does structural 'efficiency' have any relevance in this architecture; in contemporary terms, the temples have massive structural redundancy. To draw curves and asymptotical geometries in space and structures as the Greek builders were able to do, and within those complex matrices to accommodate such variation of sculptural expression, requires as much joy, vivid imagination and individual expressive intent as it requires technical mastery.

Their significance to architects working now and the original meaning of the three Greek orders are two different issues. In the history of architecture, the importance of these relics arises from their documentation and interpretation by Vitruvius and, latterly, the interpretation of Vitruvius by generations of philosophers, scholars and designers, rather than from first-hand experience of the archaeological sites. We can never see them as they were built, but if one could look, as it were, over the heads of the many interpreters with their own agendas, to the actual relics of Hellenic Greece, we should be inspired not to copy but to express the limits of *our own* imagination.

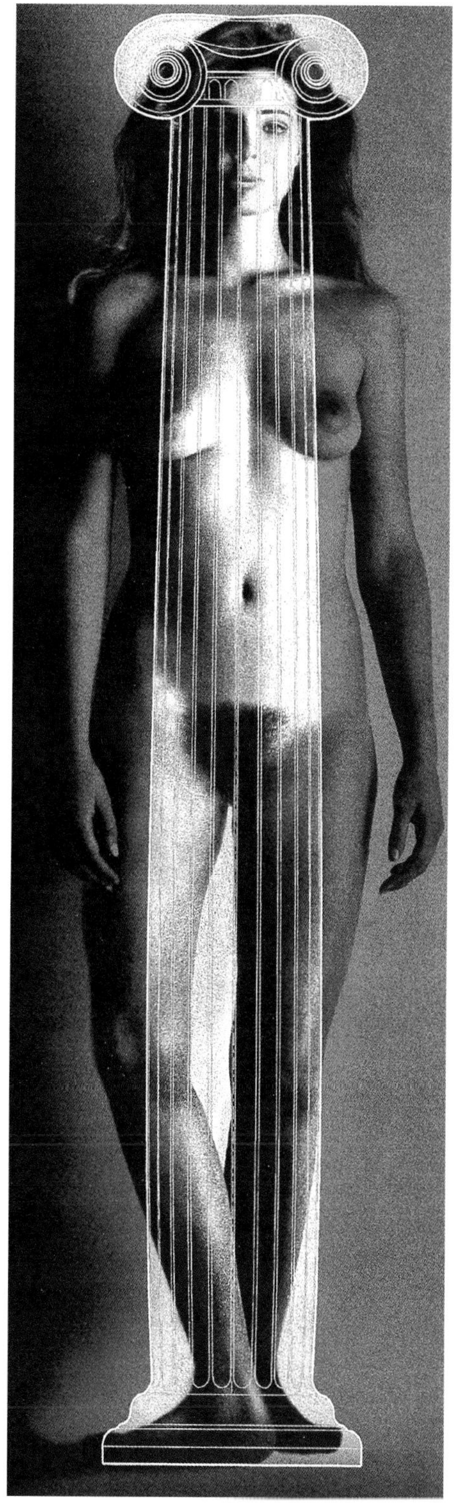

23 The Ionic order and the female form

CHAPTER 2

COLUMNS, BUILDINGS AND THE RISE OF TECHNOLOGY

24 Karl Friedrich Schinkel, memorial to Frederick the Great, *c*1840

25 Medieval columns illustrated in John Ruskin's *The Seven Lamps of Architecture*, 1849

STATUS QUO

The classical idiom remained the paradigm for architects from the Renaissance until the height of the Industrial Revolution. Many remarkable designers flowered during this technical interregnum. The practice of architecture, based as it was so securely on the classical canon and the influence of Beaux-Arts teaching, established a level of esteem close to the higher ground of the fine arts, to be endlessly picked over, manipulated and embroidered until, by the mid-nineteenth century, it had degenerated and diversified into extreme baroque, monumental fantasy and neo-gothicism: a fulsome mixture against which the work of a lonely purist like Karl Friedrich Schinkel (1781–1841) appears sparkling and fresh (24).

Towards the end of this period, endless arguments over definitions of beauty and ugliness threatened to recast the profession as a refuge for fops and dilettante connoisseurs. In Britain, in particular, the period saw a revival of interest in medieval building; for the first time in centuries it became acceptable, although perhaps a little daring, to suggest that the Dark Ages were not entirely uncivilised. The doyen of Victorian architectural writing, John Ruskin (1819–1900), was one of the gothic revival's most ardent advocates (25). In 1849, writing in his hugely popular book *The Seven Lamps of Architecture*, Ruskin analysed a row of medieval columns in the following characteristic terms:

> the vegetation growing up the two columns is fine, though bizarre; the distorted pillars beside it suggest images of less agreeable character; the serpentine arrangements founded on the usual Byzantine double knot are generally graceful but I was puzzled to account for the excessively ugly type of pillar.

Ruskin also neatly summarised the *status quo* of early nineteenth-century architecture: 'We want no new style . . . the forms of architecture already known to us are good enough.' In Ruskin's writing – and the opinion of professionals – issues of technique or detail were considered the lowly concern of builders and artisans whose own methods were unchanged and if anything (because of the sheer volume of building in the urban explosion of the Victorian era) generally less well executed than those of similar tradesmen in the centuries before.

CONVERGENCE

The revolution, when it came, was not an overnight affair. Engineers had enjoyed an heroic status since the Industrial Revolution of the eighteenth century. The riches of coal and iron were exploited by accelerating mechanisation, steam energy and the advent of greatly improved refinement techniques, to produce Portland cement, high-quality cast iron, low-cost glass and so on. Mechanisation on site permitted the construction of staggeringly audacious infrastructure projects within a few years rather than generations. The focus of creative effort was on transport (both land and sea), bridges, tunnels and utilities (26). The inventors and engineers, more often than not emerging from the artisan classes, attained a popular significance that went far beyond the influence of the architectural élite. Despite this, until the mid-nineteenth century the two disciplines were, by common understanding, strictly contrasting in their purposes.

26 William Williams, *The Cast Iron Bridge Near Coalbrookdale*, 1780

Isambard Kingdom Brunel (1806–59) is the outstanding engineer of this (and any other) period. As well as being a prodigious constructor, he was the first on either side of the architect–engineer divide to propose that the two fields might benefit from some convergence. He was appointed secretary of the Royal Commission that organised the Great Exhibition of 1851 because a consensus had already been reached that, if this event was to do justice to its nation, it required something larger and more spectacular – and more rapidly delivered – than was possible by the conventions of architecture alone. It was through Brunel's support and supervision that the concept of Joseph Paxton (1801–65) – a gardener and glass-house builder, unencumbered by architectural training – was selected and ultimately executed to house the exhibition (27).

27 Joseph Paxton's Crystal Palace, London, 1850–51, under construction

The Crystal Palace was arguably responsible for the first and greatest revision of the definition of architectural practice since its inception: it was at this point that tradition finally yielded some of its influence to the engineering concerns of invention and technology. The columns of the Crystal Palace were slender cast-iron struts. Part of a system of prefabricated components (another innovation), each column was identical, functional and without ornament primarily for ease and

28 Eugène-Emmanuel Viollet-le-Duc, V-frame, *Entretiens sur l'Architecture*, 1864

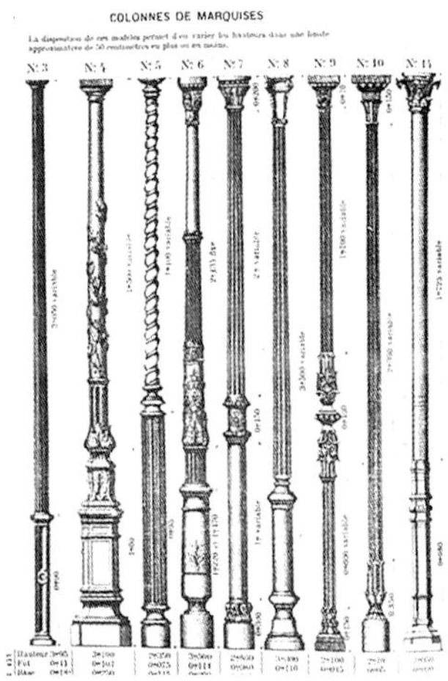

29 A page of cast-iron columns from a 19th-century catalogue of the Parisian foundry AS et Cie.

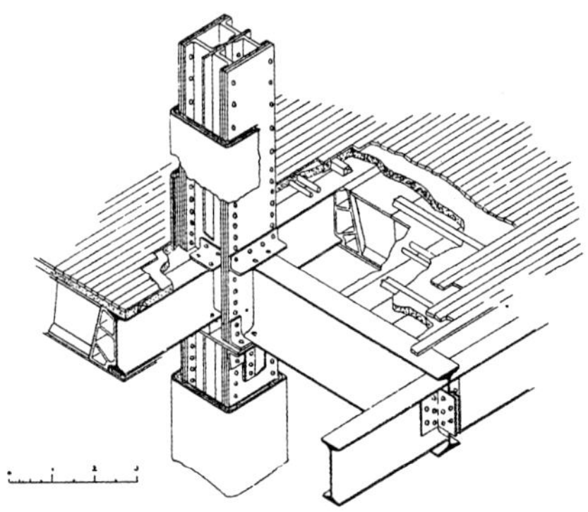

30 William Le Baron Jenney, steel skyscraper frame at column base

rapidity of manufacture. There were many hundreds of these columns in the building. The Victorian public were thrilled by the technological adventure, and frankly bored by the lack of variation in the necessarily relentless repetition of the structure. To all who witnessed it, however, it must have seemed beyond question that the art of building was transformed for ever. The first notable project to exemplify the new philosophy was Brunel's Paddington Station in London of 1851–54 (see pages 61–63).

In France at the same time, Eugène-Emmanuel Viollet-le-Duc (1814–79), another autodidact, was enthused by high gothic, but developed from it a cast-iron expressionism that was progressive, daring and highly personal. His most influential ideas were expressed in the published engravings of *Entretiens sur l'architecture* (1863-72), but generally failed to find clear expression in his executed works (28).

Late nineteenth-century society embraced innovation to an almost bizarre extent; in a sort of *fin-de-siècle* contagion, every man aspired to be an inventor. It was inevitable that invention would, at last, find its way into architecture. The British Empire opened up the world and the manifold aesthetics of its civilisations to a Western society, the swelling ranks of scientists provided all the necessary interpretation, and revolutionary transportation and communication media made sure the appetite for novelty could be fed.

By the time reinforced-concrete system building was patented in the 1870s and William Le Baron Jenney of Chicago had built the first large-scale pure steel-frame structure (Home Insurance Company, Chicago, 1883), the engineers were well in control of the future direction of architecture (30). Where some concession to classical forms had still been apparent in the columns designed for the Crystal Palace, the Galerie des Machines at the Paris Expo of 1889 – designed by the architect/engineer partnership of Charles Louis Ferdinand Dutert and Victor Contamin – was devoid of any vestigial reference (31). The Galerie's squat planar columns supported massive wrought-steel arches and were balanced *en point* on movable pin joints, in complete disregard of the convention of the pedestal base. A year later, the engineer–architects Holabird and Roche of Chicago completed the Tacoma Building, the first all steel-frame building of more than twenty storeys.

Western architecture turned the century in unprecedented freedom and diversity. From the outposts of Europe to the northern United States, architects rediscovered inventive craftsmanship, an appetite for refined, graphic ornament and a vocabulary of form that celebrated structural ingenuity and modernity. From our standpoint this is a period of transition. The Great War was looming on the horizon, and everything was set to change once again. But for architects of the period, the Chicago School, the Wiener Werkstätte, the Jugendstil, Art Nouveau, Arts and Crafts all represented a final break with classical tradition.

Art Nouveau and its related idioms barely survived the Great War. Like many other innovations of the pre-war years, from the post-war perspective it seemed tainted with the decadence of a declining Western culture. In the aftermath of a catastrophe made possible by technology, it fell naturally to the artistic community to prescribe a progressive form for society. This was the heyday of artistic movements. In music, literature and in all forms of fine art, the Modernist avant-garde claimed the aesthetic high ground.

Where artists lead the way, eventually, and with a good deal less *élan*, architects will generally follow. Charles-Édouard Jeanneret (1887–1966) was an amateur cubist painter, and precocious young architect who assumed, amongst other avant-garde practices, the habit of making declamatory statements under an adopted name: Le Corbusier. Architects never make good politicians. The polemic of Le Corbusier is restrained by the standards of the early 1920s, but it sounds shrill to contemporary ears. In his celebrated manifesto *Vers une architecture* (1923) he writes without embarrassment on classical architecture, 'emotion is born of unity of aim, of that unperturbed resolution that wrought its marble with the firm intention of achieving all that is most pure, most clarified, most economical.' Elsewhere he trumpets the ascendancy of mathematics and engineering and pours scorn on the 'peasant's' preoccupation with decoration. He claims that 'aesthetic refinement is the proper virtue of higher men', a statement which, to contemporary ears, is strongly redolent of the most acrid right-wing thinking. As radical as his intentions may have been, Le Corbusier's sociological observations are as ill-derived as his interpretation of Hellenic architecture (32).

However, notwithstanding its hyberbole and its fallacies, by its sheer verve and immaculate timing, Le Corbusier's manifesto won architectural disciples all around the world, uniting the profession under a common aim for the first time since the domination of Palladio. It is impossible to overstate the extent to which Le Corbusier's architecture recast the paradigms of building design and social accountability for architects. Since Le Corbusier, if we wish to design buildings, it seems that we are obliged to accept our share of an undeserved moral burden, and a measure of responsibility for all sorts of social ills, of which we are reminded by continual brow-beating. But Corbusier's precedent makes these burdens tolerable, if we allow it to, by allowing us freedom to express ourselves unconstrained by ancient conventions and notions of aesthetic propriety. Indeed, Le Corbusier is entitled to the credit, if not for having invented, then certainly for having popularised at least two distinct new column forms (see pages 31–34)

Before concluding this sketch of the transition from convention to invention in column design, mention should be made of the influence of Adolf Loos (1870–1933). Loos' entry for the 1923 competition for the *Chicago Tribune* newspaper headquarters is such a curiosity that it rarely escapes mention in even the most

31 Charles Louis Ferdinand Dutert and Victor Contamin, Galerie des Machines, International Exhibition, Paris, 1889

32 Le Corbusier, sketch of the Parthenon, *Vers une architecture*, 1923

33 Adolf Loos, competition entry for the *Chicago Tribune* tower, 1923

cursory account of twentieth-century architecture, despite its aberrant nature. It has to be considered here because it exemplifies the revived association of the classical orders with the moral 'purity' of the earliest Modernist functionalism (33).

Loos is known for having been an intuitive ascetic. In 1908 he published a pamphlet entitled *Ornament and Crime* which Viennese society, still balking at the daring of Art Nouveau, regarded as baffling rather than subversive. Loos' more famous projects are, as his polemic would suggest, strictly orthogonal and devoid of even the mildest geometrical inflections. But the *Chicago Tribune* tower was nothing less than a gargantuan Doric column, complete with massive pedestal, capital and a fluted shaft, perforated with hundreds of windows.

In the late 1970s and early 1980s, proponents of 'Post-Modern Neo-Classicism' often claimed to recognise a kindred theatrical frivolity in Loos' proposal. From what we know of Loos, however, we can be sure that his proposal was made in deadly earnest. Had Loos not made this proposal more attention may have been given to his offices at Michaelerplatz, Vienna (1910), and to other built works of his in which classical columns make an unashamed appearance. Perhaps also more attention would be paid to the actual content of his *Ornament and Crime* pamphlet which, far from rejecting classical forms, concentrates its criticism on the indigestible efflorescence of nineteenth-century baroque.

As far as Loos was concerned, the 'true' classical orders were above criticism, and their use in stone columns not at all at odds with Modernist functional minimalism. Such was the unassailable perfection of the Doric column that Loos could think of no monument as worthy of representing the highest ideals of modern man (with which the *Chicago Tribune* liked popularly to associate itself). Loos earnestly believed that his column tower would be a touchstone for all future architecture. He regarded the project as one that had to be built for the good of civilisation, *if not by him then by another architect.*

Loos may have been thinking of Louis Sullivan (1856–1924) when he made this concession. The firm of Sullivan and Adler dominated the skyscraper industry of post-war Chicago, and Sullivan himself had been known to make allusions to classical columns in prescribing the essential elements of a skyscraper. Sullivan believed that, like columns, skyscrapers had to have a base, a shaft and a capital, and broadly speaking his firm's towers followed this programme. However, other than in their ornate foyer areas, the structural columns of early Chicago skyscrapers, and of the great majority of later buildings of this type, are deliberately understated. More often than not, a steel universal section is cased in cementitious, timber or stone sheathing, serving to dull the bell-like resonance of the frame and to protect it from fire. Commercial space – generally the purpose of the skyscraper – puts pressure on the very existence of columns, embedding them into the perimeter and core zones.

COLUMNS AND TOWERS

Sullivan's analogy between the tower and the column invites a distinction to be made, for each form is of entirely different provenance. Where columns have mythical origins, towers, by contrast, are of earthly descent. They have evolved from defensible positions into fortifications. They have come to symbolise domination and martial power. Their *habitability* is a defining feature: a tower is a container of people or things. Loos' Chicago Tribune Tower is, for example, very definitely not a column because its habitability has primacy over any function associated with its basic verticality. Symbolism is not a 'function' in the modern sense. A lighthouse may guide the mariner, but this is incidental to the elevation of the sacred light: it is no more functional than a crucifix; the symbol would stand regardless of the distant reckoning of the mariner.

However, in certain circumstances – such as those of the Tuscan town of San Gimignano in the thirteenth to fourteenth centuries – hybrids can emerge (34). Here the towers were used to express the rivalry between neighbouring landowners; the competitive adding of height superseded any martial or accommodative purpose. Because their utility is of no significance, and their verticality of overwhelming importance, it is appropriate that they be thought of as columns.

Most columns are solid and therefore, obviously, good for no other purpose than standing and supporting, or just standing. In Chapter Four there is a discussion of monolithic columns which often incorporate basic means of ascent and occasionally, as in lighthouses, some Spartan living space. In these structures, the human occupants or other contents remain fundamentally incidental, and this characteristic, along with the manifestation of base, shaft and capital – something at the bottom, in the middle and at the top – are the defining features of a column.

34 Competitive tower building, San Gimignano, Italy, 13th–14th century

CHAPTER 3

SOME TWENTIETH-CENTURY PARADIGMS

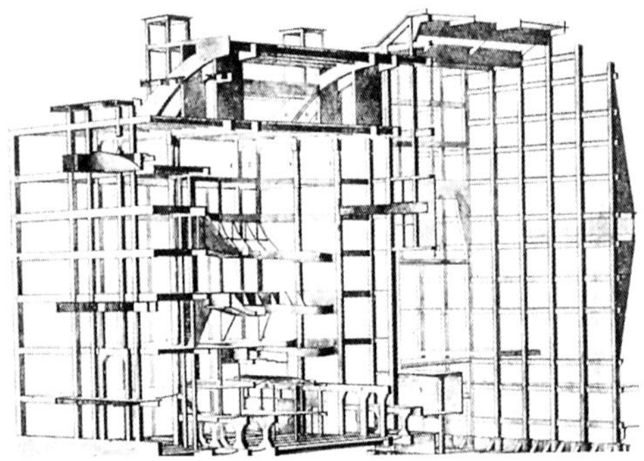

35 Auguste Perret, concrete-frame system, the Théâtre des Champs-Elysées, Paris, 1914

The delicacy, formal freedom and economic manufacture of cast iron ultimately could not compensate for its brittle nature. By the early 1900s it was fading into structural history. Reinforced concrete, a slightly later innovation, has the strongest claim to be the architectural material of this century. With steel reinforcement concrete becomes almost the perfect building material, capable of restraining the tensions and lateral loads that previously added mass in orders of magnitude to even the most lethargic masonry stump. Auguste Perret (1874–1954) was the first designer to develop an aesthetic exploiting the strengths of reinforced concrete (35), and the same capabilities are positively celebrated in several of the later column forms explored in this chapter. Since there is no obvious precedence among these paradigms, they are presented in order of their designers' birth dates. Although not necessarily original, each of them represents a model approach to the design of columns which continues to influence contemporary practice.

ANTONI GAUDÍ

Park Güell, Barcelona (1900–14)

The years when Gaudí (1852–1926) worked on the Park Güell coincided with the domination of Art Nouveau in Europe; the Park was commissioned in 1900 and built between 1903 and 1914. In so far as Gaudí can be placed in any category, Art Nouveau seems to be the source of some of his inspiration, though his individual expressionism exceeds by far any other achievement of the genre. As in Frank Lloyd Wright's architecture (see pages 28–30), Art Nouveau provides a language of graphic ornament which is seldom more than superficial in the built work.

Park Güell was built on barren, sloping ground on the northern edge of Barcelona. It was intended to be a utopian suburban colony, a harmonious enclave providing places of work, homes, allotments and the necessary urban amenities for a self-contained township. The utopian vision was not realised, though some spectacular structures were built, and these, along with the exotic planting, have survived to form one of the city's most popular communal spaces (36).

The highest level of the proposed township was an open plaza, a market place. Alongside it was an open-air performance area which Gaudí imagined might address the natural auditorium of the adjacent parkland. Both of these open platforms hang out over the planted slopes of the park. The spaces below the projecting decks are filled with columns. The space below the 'Greek theatre' is an entrance lobby filled with columns of a reasonably faithful Doric pattern, arranged in a dense, irregular hypostyle (37). Gaudí's devoted patron in this and many of his later projects, the industrialist Eusebio Güell, was an educated man with a passion for classical history. The unexpected manifestation of the Doric order here is probably a concession to his client's enthusiasm – a precautionary invocation of the mythical power of the Doric order to exercise a spiritually edifying effect upon the residents of the new utopian community.

The outermost row of Doric columns leans back towards the crest of the plateau at a precipitous angle. This establishes the geometry for the promenade columns which support the projecting edge of the plaza. These columns are without classical features, inventive, organic and technically ambitious. They are inclined at various angles, roughly 30 degrees off the vertical in the most extreme cases, and built up in helically wound courses of hand-made bricks and rusticated basalt. The diameters of the columns vary between 30 and 60 centimetres, for a vertical span in the order of 4 metres, although in some areas, as the hillside drops away, the columns are double-banked to create a two-storey promenade.

At the capital, the columns flare out into whirlpool cones, some of which contain seating and some of which are filled with shrubs. The perimeters of the cones merge to form a continuous rocky balustrade, overflowing with the foliage that is a critical component of the architecture.

The organic, rustic appearance of the columns can mislead one into thinking that their construction is somehow accidental. Gaudí made a huge investment in structural modelling and testing in order to achieve the 'naturally grown' quality. His method of structural design, intuitively realised, is identical to that 'pioneered' over half a century later by the great engineer Frei Otto (b1925). Gaudí would hang a multitude of carefully weighed and positioned bags of sand from a network of

36 Antoni Gaudí, Park Güell, Barcelona, 1900–14: helical brickwork columns of the promenade

37 Doric-type columns in the entrance lobby beneath the plaza, Park Güell

hanging cords; each of the bags represented the lines of force acting on the cord, and the profile traced from the cord, when inverted, would provide the angles for setting out the columns. By this means he could be sure that the angle of each column was perpendicular, at all times, to the load it carried. In the case of the Güell family chapel, which was built during the same period as the park, the process is used in a more clearly confined structural system (38).

38 Antoni Gaudí, Güell family chapel, Barcelona, 1905

The execution of the columns required shoring and formwork of the kind conventionally used in the construction of brick arches and vaults, and placed the highest demands on the skills of the bricklayers. In every respect, even by contemporary standards, these structures are marvels of advanced structural design and building method, in which the technology remains subservient to the expression of a distinct naturalistic imagination.

FRANK LLOYD WRIGHT

Johnson Wax Administration Building, Racine, Wisconsin, USA (1927–39)

If anything more revolutionary in his vision than his European contemporaries, the great Welsh-American architect Frank Lloyd Wright (1869–1959) produced work that is characterised by an organic homogeneity that resists conventional elemental analysis. Whereas Le Corbusier's buildings follow the oldest conventions of architecture in that they are designed by the hierarchical arrangement of formal components, allowing one, for example, to look at the columns as if in isolation from the whole, Wright's buildings typically have an organic character that makes it difficult, or at least unenlightening, to consider any part discrete.

39 Frank Lloyd Wright, Johnson Wax Administration Building, Racine, Wisconsin, 1927–39

Perhaps because America largely escaped the cathartic experience of the Great War, Art Nouveau did not fall so suddenly into disfavour as it did in Europe. Its influence was evolved-out-of rather than abandoned, and it lends a gradually diminishing but distinctive colour to Wright's designs throughout his long career. Art Nouveau is undoubtedly one of the sources of the naturalistic approach advocated by Wright which, in the 1930s and 1940s, seemed so profoundly at odds with the new European conventions. Wright cultivated the contrast in approaches, and hoped that his way would be recognised as distinctly American.

Instead, this quality is regarded today as distinctively 'Wrightian', because few, if any, of those designers who have tried to work by his method have had the exceptional skill and control that it demands. Hierarchical composition remains the convention for most designers because it is so much easier to apply.

The columns for the Johnson Wax Administration Building in Racine, Wisconsin, encapsulate Wright's method: they are columns but each of them is also integrally a roof and a ring beam (39, 40).

Variously described as 'dendriform' (tree-shaped) and 'lily pads', the columns really bear little resemblance to either. Wright was known to have claimed that Minoan column forms were a precedent for their design, although he may have been making mischief here by alluding to an archetype that predates the classical period. While they are certainly wider at the top than at their base, they have little else in common with Minoan forms.

Where Le Corbusier's preferred medium was *in-situ* cast concrete, often bearing the texture of the timber formwork (the finish known as *beton brut*), Wright made inventive use of pre-cast concrete. His principal building material was concrete block, purpose made, occasionally on fine mesh reinforcement, in pattern moulds of his own design. This was technologically advanced for the time, and when Wright proposed that the columns for Johnson Wax be precast in light-gauge mesh-reinforced concrete, he and his collaborators probably knew more about the technology than anyone else in the world. Re-usable polished steel moulds were utilised for the job. Using the techniques developed to manufacture his patterned concrete blocks, Wright embossed the mould to give leaf-like patterns to the collar of the shaft and the underside of the dish-shaped capital.

The columns are hollow for most of their height, the concrete never more than a few centimetres thick. At less than 25 centimetres in diameter at the base, they flare out, over their 7-metre height, to a diameter of around 5 metres at the crown, allowing adjacent columns almost to touch at the rim and obviating altogether the need for substantial roof beams.

This is not a frequently used column form: a lofty space is required to make sense of the extreme flaring of the column. It resolves classical tree symbolism into more direct metaphor, augmenting the conventional shaft – or trunk – with a spreading arborescent crown. The apparent 'flowering' at the head of the columns calls to mind another singular structural assemblage: the gothic vaulted nave (41). The most remarkable recent manifestation of the form was in Tadao Ando's Japanese Pavilion for the 1992 Seville Universal Expo, a building that was basically a lightweight timber skin wrapped around ten towering skeletal dendriforms (42).

The Johnson Wax columns were designed in 1937. Each was expected to carry a load of around 6 tonnes. On the assumption that any such column would be of masonry construction, Wisconsin building codes demanded that they should have a minimum diameter of 75 centimetres at the base. The celebrated 'proof' of Wright's columns, before invited pressmen, involved the gradual loading of a sample with sandbags until the design load was attained. Having demonstrated its complicity, Wright then ordered the load to be increased until, by stages, it reached 60 tonnes – ten times its design load – at which point the timber shoring keeping the sample upright itself collapsed. Although the 'capital' snapped on impact, the shaft of the column survived the fall (43).

40 Detail drawing of a column in the Johnson Wax Administration Building

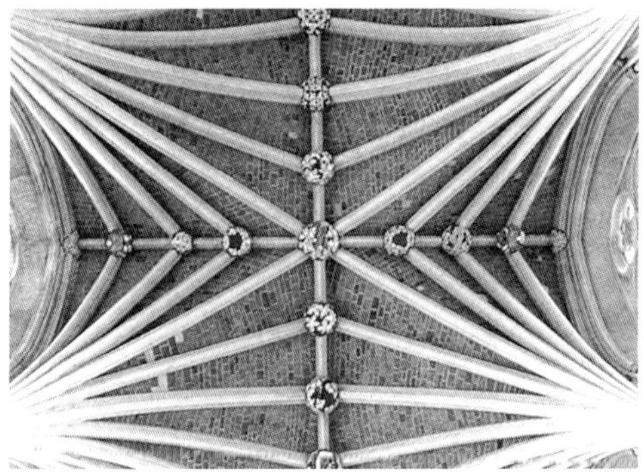

41 Gothic fan vaulting, Exeter Cathedral

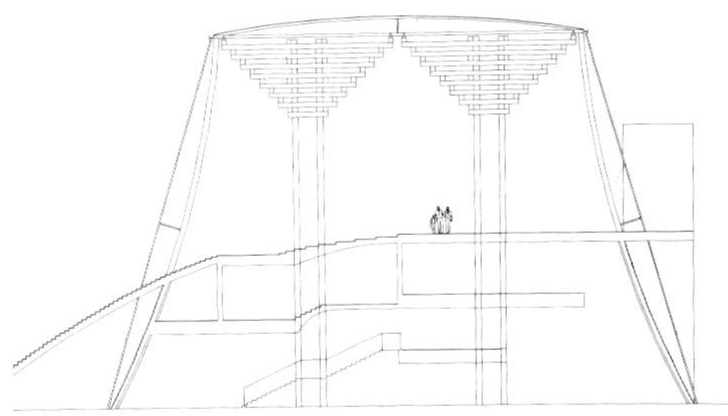

42 Tadao Ando, Japanese Pavilion, Seville Universal Expo, Spain, 1992

43 Testing the load-bearing potential of Wright's columns for the Johnson Wax Administration Building

For all the rigour and aesthetic refinement that Le Corbusier expressed in reinforced concrete, it was Wright more than any of his contemporaries who should be credited with first realising something of the full range of its expressive potential. Le Corbusier's canon is so thoroughly assimilated into Western architecture that it has itself become a source of well-worked formulae; the classical canon of the twentieth-century. Wright's architecture is more diverse, more expressive and more personal; and these are the qualities which, at the present time, it appears architecture needs to cultivate in order to continue to contribute usefully to society.

LUDWIG MIES VAN DER ROHE

New National Gallery, Berlin (1961–68)

I do not wish to be interesting; I wish to be good.

Mies van der Rohe

44 Mies van der Rohe, New National Gallery, Berlin, 1961–68

In Mies van der Rohe's terms, 'good' was explicitly associated with a flagellant poverty of emotion. He came to share with the devotees of certain extreme religious orders the bizarre notion that the deprivation of asceticism, in his case in architecture, was a means of achieving beatification.

Like all cults, the cult of minimalism offered seductively easy answers in exchange for the sacrifice of individual expression; in the reformative years that followed the Second World War, a great many architects were won over to the Miesian way, and lasting public hostility to the profession has been the result.

The new National Gallery in Berlin was the last great work to be completed in Mies' lifetime (1886–1969); it is truly the apotheosis of his craft and, despite its determined lack of overt expression, it bears the unmistakable signature of the 'master' (44).

Mies was apparently ill at ease with the sloppiness of concrete in his mature work, and even balked at the use of hollow steel sections for his columns, presumably because they could not be 'exposed' in their entirety. For the National Gallery complex, steel fabrications are used to accommodate the same stress distribution that much smaller hollow sections could have contained. The equilateral cruciform plan of the columns can be read as a 1-metre hollow square section with its corners cut away, and with substantial cross webbing installed to compensate for their omission (45). This is an extreme, almost irrational mannerism disguised as a minimal absolute.

The roof of the gallery is a fabricated steel two-way beam system over two metres deep, which spans clear across the square, column-free box of the gallery. There are only eight columns to support the entire roof which, at around 60 metres square, was the largest rigid plate ever made at the time of its construction. This massive slab actually sits on bearing joints at the head of each column. These allow the considerable expansion movement of the roof to be accommodated without bending the column. Through each of the discreet, rotation-free pin joints flows a force of many hundreds of tonnes.

Also shown here, for comparison, is the plan of a column from the legendary German Pavilion at the Barcelona Expo, 1929 (46). Forty years separate these two projects, but in conceptual terms they are identical. Both buildings elevate flat roof decks on eight cruciform columns. The 1929 column is less refined only in so far as it is coated in chromed steel plate. Mies' notorious claim that 'God is in the details' may be a glib play on words, but it is also an echo of Renaissance faith in the origin of the Greek orders – access to the sublime achieved through minute observation of an inflexible formal grammar. Mies clearly believed himself to be on a similar spiritual mission, and over forty years the unchanging, obsessive focus of the truly devout informs the austerity of his vision.

Despite its undoubted brilliance and the unprecedented control exercised in its realisation, like most of Mies' later work the building offers little to engage the uninitiated. The planar simplicity of the columns, eschewing even the ambiguity of volume, belies the self-indulgent complexity of their manufacture. The columns again provide a satisfactory microcosm of the whole: a building in which 'classicism', 'purity' and 'God' collide with chilling effect.

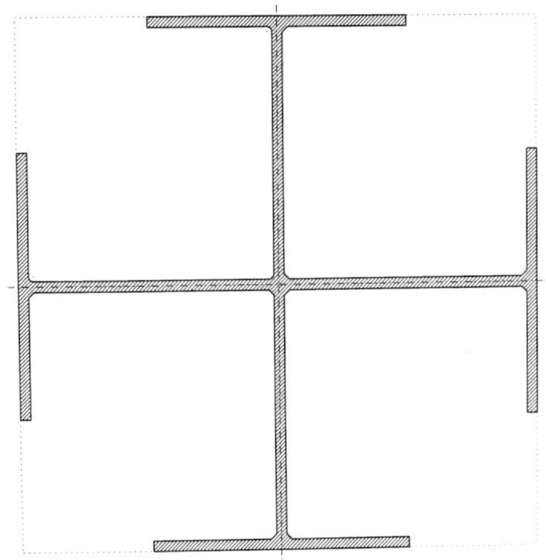

45 Mies van der Rohe, plan for a column in the New National Gallery

46 Mies van der Rohe, plan for a column in the German Pavilion, Barcelona, 1929

LE CORBUSIER

Le Corbusier (1887–1966) was extraordinarily inventive. Where previously architecture had been hidebound by the idea of emulation, he can probably be credited with adding inventiveness to the necessary virtues of the practising architect. There are few of his noteworthy innovations that do not still haunt the practice of architecture. His influence is comparable to that of Picasso on twentieth-century art: it is impossible to imagine how creative expression might have developed without his influence, and impossible to imagine that such a vast new vocabulary could ever again be made accessible to the art form.

If any element of a building can represent the whole in microcosm, it is often the column. In Le Corbusier's canon, the architect's fidelity to the philosophy of classicism led him deliberately to invest his columns with the greatest significance. The following examples represent key stages in Le Corbusier's development, and each also represents an addition to the lexicon of basic column forms.

47 Le Corbusier, Villa Savoye, Poissy, France, 1928–30

48 Le Corbusier, Palace of Assembly, Chandigarh, India (completed 1964)

49 James Stirling, Olivetti Headquarters project, 1971

Villa Savoye, Poissy, France (1929–31)

Le Corbusier used various devices to express his sympathy with classicism. One such mechanism was the evocation of the natural state or resonance of the materials. Modern aesthetics recognise in the antique relics the isolated beauty of weathered stone. In the Villa Savoye it is the slender, widely spaced columns that resonate at what feels like their natural pitch: *slender* and *widely spaced* (47). Their profile and positioning emphasise their own tenacity as well as that of the beams spanning them. Together they could be said to imply that the *structural capability* of the material – reinforced concrete – itself merits expression. In the case of the Crystal Palace (see page 21), the building could not have been achieved without the innovative use of cast-iron columns, and the technology was the means of achieving this end. In the Villa Savoye, there is no explicit requirement for the exaggerated emaciation of the concrete structure: its structural capability is expressed for its own sake. This idea had not been a significant motivation in Western building since the twilight of medieval gothic masonry.

Le Corbusier appears not to have been attracted by what many architects who followed him regarded as the logical conclusion of structural reductivism: the fantasy of limitless column-free enclosure. For Le Corbusier, the cultural importance and expressive potential of columns made them far too valuable ever consciously to omit. The columns of the Villa Savoye are part of a formal composition designed to express the elevation above the ground of the apparently heavy upper storeys. The slenderness of the columns suggests that the upper portion of the building is, despite appearances, almost weightless.

This compositional device has been hugely influential in twentieth-century design. The ultimate liberation of building from the ground-plane became a key feature of the Modernist utopian vision. Le Corbusier himself made something of a speciality of it, as the following examples attest. The Villa Savoye is theatrical rather than idealistic: the lightness of the elevated structure is an illusion achieved by the use of a grid of slender columns. If the appearance of lightness is an objective in a composition of this kind, it is always more effective to use a large number of very slender, perhaps randomly positioned columns than it is to aim for the widest possible spans between heavier supports.

Chandigarh, India (1950–65)

The new regional capital of the Punjab, Chandigarh, allowed Le Corbusier a rare opportunity to exercise his formal invention on an epic scale, without constraint. He worked on this project from 1950, when he was commissioned to prepare the master plan, until his death in 1965. Within this collection of civic buildings and monuments, he introduced several refined column archetypes which, as always in his work, are deployed for maximum dramatic effect.

The forum of the Palace of Assembly (completed in 1964) is a double/triple-height hypostyle hall which transposes the athletic leanness of the classical, externally expressed columnar matrix, as seen at the Villa Savoye, to an enclosed internal volume, with similar elevating effect (48). Whereas the masonry hypostyles of pre-modern architecture are typically infused with such a sense of density that one might imagine the whole weight of the heavens crushing down overhead, those of the Palace of Assembly, at 38 metres in height, convey a sense of tension: it is almost as if they are tethers rather than props. The conical 'capitals' provide a cosmetic termination; the structure continues above the plane of the ceiling.

This column form has been appropriated many times. James Stirling's project for the Olivetti Headquarters (1971) is just one of many undisguised homages (49). The columns of the factory floor of Sir Norman Foster's Thames-side headquarters in London (1993) have the same provenance.

Chandigarh is noteworthy, also, for two monolithic column forms. The first of these is the mis-named 'dome', a hyperboloid cylinder unashamedly inspired by the cooling towers of a power station, which comprises the lower chamber of the Palace of Assembly. This hollow column is designed, in the finest primordial tradition, to capture diffuse beams of sunlight, heightening the chiaroscuro drama of the forum space. The second of these monolithic forms is the Monument of the Open Hand of 1952 (50). A thoroughbred totemic column, its (perhaps exaggerated) sense of its own symbolic potency is somewhat undermined by a suggestion, in the form of the base, and the outstretched hand signal above, that it could be a giant bus stop.

Notwithstanding the occasional pomposity, this project and its contemporary, the extraordinary monastery of La Tourette near Lyons, France (1957–60), with its towering undercroft and polyrhythmic column/mullions, are two of the great endeavours of creative building in the twentieth century (51).

Another roughly contemporary project, the Ahmedabad Cultural Centre (India, 1956) offers a convenient, less admirable counterpoint. It is, however, the provider of another prominent column archetype. The mass of this building is, again, raised on a grid of simple cylinders (52). The elevated facade, blank and unperforated, is approximately four times the height of the columns beneath, and the height and girth of the supports, in this case, emphasise and exaggerate the mass of the elevated structure. There is no tension or elegance. There is, however, the guaranteed problem of a vicious under-draught. In so far as it relates to the use of columns, it is worth repeating here that a combination of a low, open undercroft below a tall flat fascia is the best possible means of amplifying even a gentle breeze into a gale at ground level. Unless this effect is, for some reason, desired, this combination of forms is best avoided.

50 Le Corbusier, Monument of the Open Hand, Chandigarh, India, 1952

51 Le Corbusier, Monastery of La Tourette, Lyons, France (1957–60)

52 Le Corbusier, Cultural Centre, Ahmedabad, India, 1956

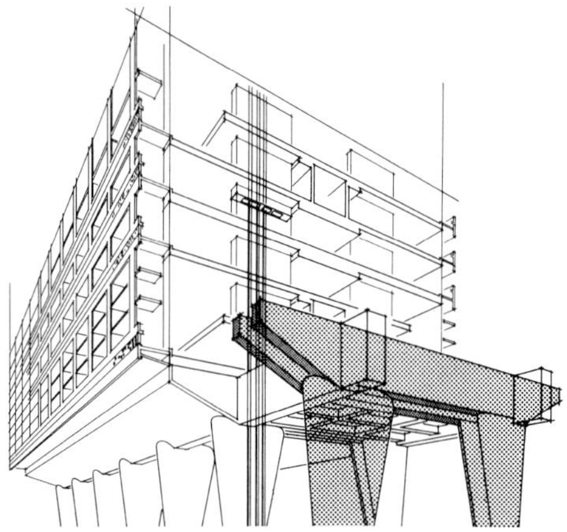

53 Le Corbusier, Unité d'Habitation, Marseilles, France, 1947–52

54 Le Corbusier, Unité d'Habitation

Unité d'Habitation, Marseilles (1947–52)

The wind tunnel effect is also in evidence, although restrained (thanks to the efforts of residents over the years to screen the undercroft), below the massive apartment building, the Unité d'Habitation (53, 54). In chronological terms a predecessor to the Indian projects, this building is the most familiar example of a transitional column form in which the supports take on the characteristics of exposed structural walls.

Twenty years earlier, in the exquisite Pavillon Suisse in Paris (1930–32), Le Corbusier had been compelled to abandon his first scheme because the four tiny steel cylinders upon which he proposed to elevate the accommodation were demonstrably inadequate for the task. They grew, in the executed project, into a row of concrete slabs orientated edge-on to the front elevation so that, in drawings at least, they would still appear stalk-like. In reality, though, the supports are never all seen edge-on. Le Corbusier came to accept that the mass of the supports, when called upon to carry such huge loads, would have to be accommodated and if possible given a distinct and non-columnar expression.

While the supports of the Pavillon Suisse still strive for the pared-down appearance which predominates in Le Corbusier's earlier work, and which is usually taken to reflect his sympathy for Hellenic building, those of the Unité d'Habitation are invested with more than their basic elemental qualities and are well on the way to becoming something else. They are re-cast (literally) as two rows of organic conical vessels. The evocation of tension/compression is ambivalent and anyway subsumed by the anthropomorphism of the sculpted forms.

The columns are also hollow rising ducts for all the services into and out of the massive apartment block above. For this reason, they are U-shaped in plan, arranged in splayed opposing pairs, like legs, with neat cubby-hole doors on the hidden side. More importantly, they have, cast into their surfaces, the recessed silhouette of Le Corbusier's Modulor Man. It is this device that most emphatically conveys the persistence of classical ideas in the architect's creativity – an idea of the necessity of a link between building and human proportion which echoes Vitruvius' decoding of the orders. Whereas there is no overt classical style in his building, Le Corbusier evidently believed that through his own magical proportional theory, from which the design of this and many later buildings derives, he could achieve the spiritual transcendence which had once been within the reach of architects.

Whereas the proportional theory failed to catch on with other designers, the image of the great, anthropomorphic *piloti* was rapidly assimilated into the architectural vocabulary, to reappear in a thousand buildings in hundreds of cities and towns around the world, but never since with the beauty of the original.

ALVAR AALTO

Villa Mairea, Noormarkku, Finland (1938–39)

By the time Aalto (1898–1976) was commissioned to design the exquisite Villa Mairea, he had already lost faith in the mainstream of Modernist functionalism. In the 1920s he had ventured into experimental mass housing, but as his own aesthetic matured, he began to believe that the rigid, unvarying orthogonal grid structures that manifested the unchallenged rubric of industrialised architecture were, in fact, detrimental to a healthy environment. Aalto became convinced that the adventitious rhythms and textures that he saw in nature, particularly in the natural occurrence of trees, could provide a more subtle and more edifying geometry for building. He was probably right (55).

55 Alvar Aalto, Villa Mairea, Noormarkku, Finland, 1938–39

The columns that punctuate the fluid rectangular volumes of the open-plan living areas of the Villa Mairea are consistent with Aalto's structural philosophy in that they seem to be randomly deployed. They pass up through both floors of the house; emerging in awkwardly central locations in some of the smaller first-floor rooms where they are, however, the most welcome of intrusions.

The columns are sometimes single shafts, but are more often in pairs. Around the sauna cabin they are grouped in clusters of three or four at a time (56).

56 Triple columns around sauna cabin, Villa Mairea

Secondary staves, which form screens and balustrades, enrich the arborescent theme. Coniferous woodland presses close to the house on all sides: the columns and other verticals are designed to suggest that the house is woven right through the virgin woodland, that house and trees share the same space.

The informality of the planning, with its gradations of externality, filtration of light and heady naturalism, is powerfully redolent of traditional Japanese building. Partly wrought timber column clusters near to the sauna are tethered with cane strip and founded on unworked boulders – another characteristic Japanese detail.

The internal columns are built up around hollow steel cylinders 15 centimetres in diameter. Their cores are packed with concrete, and they are sheathed, perhaps over cautiously, with asbestos cement casings and painted gloss black. Each of the columns has an outer ornamental timber sheathing on the part most likely to be touched (that between knee- and head-height). In most cases they are spiral wound with natural rattan cane. The tight binding of variegated cane gives a warm, organic texture to the otherwise inert shafts. Where the columns are cast in pairs, the rattan is occasionally wound around both, transforming the paired cylinders into radiused slabs (57).

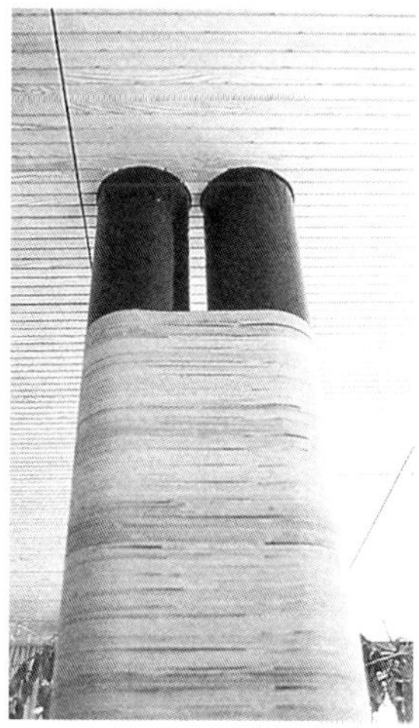

57 Double column, Villa Mairea

Columns in the first-floor rooms are sheathed in vertical, faceted beech wood strips, again over the central portion of the shaft only. This covering more obviously refers to classical fluting. These seem exceptions to the prevailing aesthetic of the house. Although, like all great architects, Aalto's sympathy with classicism has been the subject of frequent conjecture, the columns of the Villa Mairea are direct tree analogues, rather than tree symbols.

The house is a fusion of European Modernism, Scandinavian regionalism, and the philosophical repose of Japanese design; but for all the obvious and diverse inspirations, the whole has a balanced resolution. The hybrid aesthetic that we now recognise as Aalto's invention has itself been immensely influential on later design.

The column pairs have occasionally been revived, notably by Carlo Scarpa. However, the more significant column paradigm to originate with the Villa Mairea is the slender, randomly spaced rod, sometimes vertical and sometimes at an angle. This has featured frequently in the progressive architecture of the 1980s, in the work of OMA, Enric Miralles and William Alsop among a host of others, and is now well assimilated into mainstream design.

JEAN PROUVÉ

Contemporary high-tech architecture is an heroic enterprise paradoxically supported by humanistic ideals. The icons of high-tech are few and carefully selected to verify what could otherwise appear an unconvincing marriage of machismo and liberal sentiment. The three designers most frequently alluded to as progenitors are Richard Buckminster Fuller (1895–1983), Pierre Chareau (1883–1950) and Jean Prouvé (1901–84). Fuller and Prouvé in particular were global thinkers who devoted their lives to finding ways of making industrial manufacture useful to those in most need.

There is no nostalgia, romanticism or conscious image-making in the work of these designers. They were engineers, unencumbered by formal architectural training, and for this reason the forms and philosophy of their work originate outside the continuum of architectural evolution, and appear unburdened by the superimposed layers of historical reference that haunt even the most radical classic Modernism.

Fuller's Dymaxion '4D' House of 1927 was never built in its entirety, but its bathroom, which was separately prototyped and patented in 1938–40, made revolutionary use of the inherent stiffness of folded sheet metal. The greater reputation of Fuller is based on the engineering of complex 'space-frames' which achieve massive spans by the three-dimensional interlocking of repeated light-gauge metal struts. This type of structure typically does not distinguish between vertical and horizontal: columns are not required.

The celebrated Maison de Verre, Paris (1928–32), by the furniture designer Pierre Chareau, was an assemblage of radical 'industrial' references. The agenda and influence of Chareau's remarkable refurbishment project are often likened to

Prouvé's, but, unlike Prouvé, Chareau is confined to the conventional range of interior design ambition.

Jean Prouvé was originally a craftsman metallurgist. His interest in finding domestic applications for mass production in lightweight metals drew him into mass-housing design, and latterly into larger scale building experiments. Like Fuller, he was an instinctive humanist who regarded the provision of low-cost shelter as the highest goal of his craft. For this reason, the success of a designer's work could be gauged, in Prouvé's terms, by its fidelity to the maxim that 'most' should be achieved with the 'least possible means'. However, the influence of his work on later designers derives less from his theoretical objectives than it does from the considerable Gallic flair, the elegant lines, and the unintentional stylistic flourishes that make it unmistakably Prouvé.

Maison du Peuple, Paris (1937–39)

The compounding and hybridisation of building elements, to reduce weight, *in-situ* junctions, and component numbers is a characteristic of Prouvé's building systems and for this reason, as in Fuller's work, there are relatively few explicit columns. The columns that support the roof perimeter and glass walls of the Maison du Peuple in Clichy, Paris, are secondary structural elements (58, 59). The primary frame, as in the majority of his structures, is a matrix of hollow, pressed steel portal frames. The secondary structural mullions, like the main frame, are fabricated from thin sheet steel, rolled into semi-elliptical tubes. The section is hollow and pressed with integral recesses to accommodate the edges of the double-layer glass fascia panels. The glass panels are clamped into place with bolted secondary pressings.

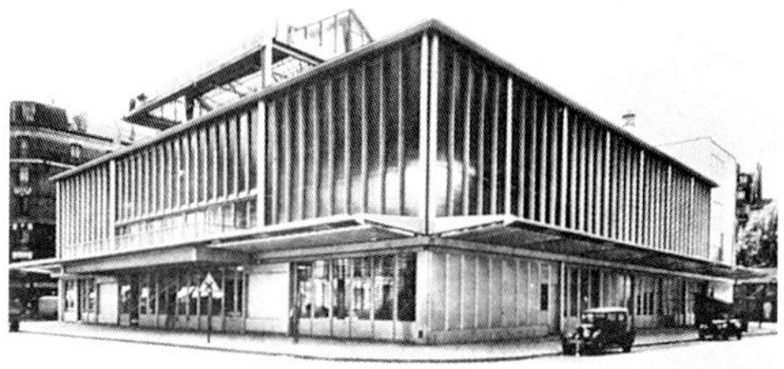

58 Jean Prouvé, Maison du Peuple, Paris, 1937–39

As commonplace as such systems may be sixty years later, at the time when Prouvé made this proposal, the wide-span double glazing and pressed-metal columns were without even tenuous precedent. Here, with a single layer of material barely 3 millimetres thick, Prouvé has been able to manufacture rigid columns standing 8 metres tall and bearing considerable wind and structural loads.

For a short while, in Europe and America, the radical technology of building was at the forefront of progressive design, feeding through to automotive design, and furniture and industrial design.

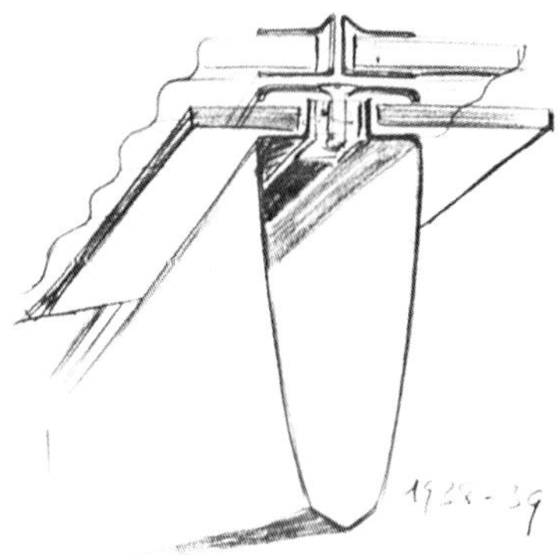

59 Detail drawing of a column in the glass walls of the Maison du Peuple

Very few of the designers who drew inspiration from the imagery of Prouvé's functional experiments possessed the scientific knowledge and inventive instinct to sustain its genuine technological endeavour. The academic architectural establishment is quite properly preoccupied with artistic expression, and most architects' interest in technology reflects the allure of its visual glamour, rather than its social utility.

RICHARD ROGERS AND RENZO PIANO

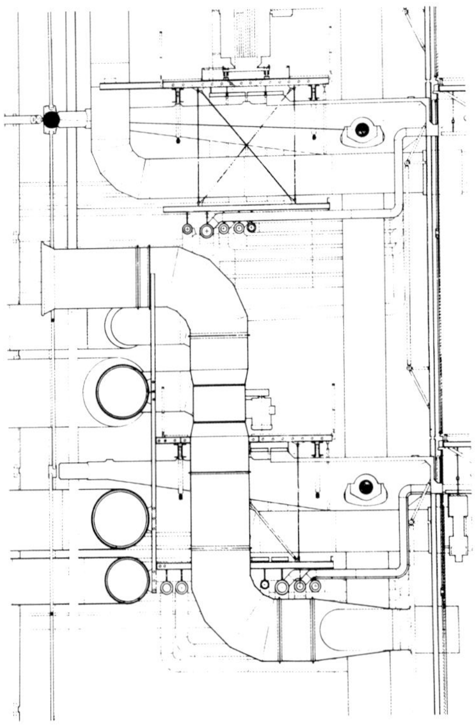

60 Richard Rogers and Renzo Piano, facade section of the Pompidou Centre, Paris, 1972–77

Pompidou Centre, Paris (1972–77)

Many of Prouvé's inventive ideas were to resurface, after decades of mute system building, in the spectacular 1972 design for the Pompidou Centre in Beaubourg, Paris (completed in 1977). The credit for this design is split along uncertain lines between the Italian Renzo Piano (b1937) and the Anglo-Italian Richard Rogers (b1933), but many 1960s architectural academics, particularly the fashionable coterie of London's Architectural Association, have claimed to have inspired it. The AA of that time was characterised by the tangle of boyish mechanistic fascinations, English neuroses, and stale hippy philosophy that were to become the unspoken themes of high-tech; but the dramatic formal language through which these otherwise innocent nostalgias find expression in building originates decades earlier with Prouvé, Chareau and Fuller and other designers of the first 'machine age'.

The evolution of the winning competition design for the Pompidou Centre into the built work is that of an effusive yet insubstantial fantasy into a piece of muscular neo-constructivism. The industrial pragmatism of Prouvé and Fuller is mutated into machine imagery and the merely implied capability of movement or flexibility. The building is arguably none of the things it set out to be. Static, monolithic, ornate and overbearingly massive, it also recalls the philosophy of the Italian Futurists – to whom the architects are certainly connected culturally – which embraces the submission of humanity beneath the juggernaut of industry. It has often been said that, when they imitated the Pompidou Centre, later architects made a formal model out of Piano and Rogers' one-off counter-culture adventure; but from our standpoint, the Pompidou Centre appears every bit the formal prototype, replete with compositional hierarchy and mythological symbols. This doesn't make it a failure. In all important senses, it is one of the most successful grand projects in European history.

The building has a primary structure which acts as a shelf rack. The absence of internal columns makes enormous demands upon the performance of the 1-metre-diameter perimeter columns, and to make matters worse, the columns are positioned outside the zone of the floor plates which, in conventional circumstances, would have braced the vertical span at regular intervals. At an average of around 50 metres in height, they appear disproportionately slight, but this befits the wiry, fuse-board imagery of the exterior (60).

Intermediate floors occur at 7.5-metre intervals, and at each level the columns are anchored to massive trusses that span the full 48-metre width of the building. The trusses are deep enough almost to occupy half of the height of each floor level, thus inevitably compromising the hoped-for flexibility of the open floor spaces. Each pair of columns supports six of these 120-tonne trusses. So much is suspended from so few verticals that catastrophic collapse might easily result from the kind of partial failure that a fire could cause. The conventional precaution would be to case the columns with concrete. But the architects had come to place such great importance on the clear exposition of the structure that the prospect of covering the primary steelwork was quite offensive to them. As a compromise, the authorities agreed to the extraordinary suggestion that the giant cylinders should be filled with water which, in a serious fire, would boil off at the head to cool the steelwork and protect it from collapse: these columns are the biggest vases in architectural history.

Although they do most of the work, the primary columns are concealed for most of their height by the network of service ducts and other paraphernalia that ornament the exterior. Some disquiet was expressed by the design team, during the long evolution of the project, that a superficial structure would have to be fixed to the exterior simply to support the exposed viscera. Like the fire protection, this was considered so deeply at odds with the tenuous rationalism of the project that some means of tying the superficial support into the primary structural system of the building came to be regarded as critical to its aesthetic integrity.

The engineers, the ubiquitous Ove Arup and Partners, allocated the project to their most able designer, Peter Rice. He is credited with having provided the structural justification that was needed for the double-layered external framework of the building. His 'solution' required that each truss should be connected to the columns through huge, pivoting steel brackets. The brackets act as see-saws with the column passing through the centre acting as the pivot. The inside end of each is attached to a floor truss, and the outside end to a grid of steel tethers that restrain the brackets from tipping under the weight of the truss. This balanced cantilever frame was a model suggested by the German engineer Heinrich Gerber (1832–1912) in the nineteenth century. The huge cast-steel brackets designed by Rice for the job have been referred to ever since as 'gerberettes'. They are, arguably, components of the primary columns because the columns' tendency to bend under the asymmetric load of the trusses is regulated by the tension in the outer grid, thus artificially widening the structurally active zone of the column. It is partly this remote stiffening that allows the primary verticals of the Pompidou Centre to be made so narrow (61, 62).

The same method of widening is frequently manifested in the mast structures of later high-tech buildings. It has occasionally been remarked that this widening is analogous to the swelling, or entasis, which is such an important characteristic of the shafts of the classical column types: this suggestion is quite erroneous. Any similarity is coincidental. The entasis of classical columns has nothing remotely to do with structural economy, a concept as alien to Hellenic architecture as mechanical escalators.

61 Placing a gerberette, designed by Peter Rice, on to a column, Pompidou Centre Paris

62 Cast-steel gerberettes in the foundry where they were produced

The technical ambition and constructional expertise invested in the Pompidou Centre's design, and the extraordinary precision and craftsmanship in its execution, does, however, justifiably bear comparison with the most brilliant architectural achievements of the past. It established a standard of excellence that has been out of the reach of the great majority of architects and which – although it still seems to be the prerogative of Rogers and, in particular, Renzo Piano – was not achieved by another architect until the emergence of the nascent English high-tech style, and the particular manner of Rogers' former partner, Norman Foster.

NORMAN FOSTER

Renault Parts Distribution Centre, Swindon (1980–83)

63 Norman Foster, Renault Parts Distribution Centre, Swindon, 1980–83

The Renault Parts Distribution Centre was the first large project by Foster (b1935) to depart from the seamless-shell aesthetic that had first brought him to the attention of large-scale corporate clients. Richard Rogers had shortly beforehand completed the Fleetguard Distribution Centre in Brittany (1979), and his overt use of masts to support the large flat roof of that building was to consolidate this formal device as a key element of the high-tech repertoire.

The mast columns of the Renault Centre are its defining feature (63, 64). The whole of the building appears subservient to them and their curious umbrella arms. Foster made a point of creating an undulating roof to follow the angles of the outstretched arms, thus distinguishing his building from Rogers' which was merely 'a flat roof with appliqué'. To ensure that the arms are indeed read as an eruption of the vertical posts, they are deployed symmetrically about each shaft regardless of whether they are holding up the roof or, as at the perimeter and in the empty bay of the *porte-cochère*, just reaching out into space. Even the internal columns are skirted with transparent panels where they penetrate the roof plane so that they remain assertively detached from the envelope of the shed. The enclosure elements, the walls and roof, seem so awkwardly crimped by their compliance with the structural lines that they appear almost to have been pushed into the gaps, as an unwelcome concession to necessity.

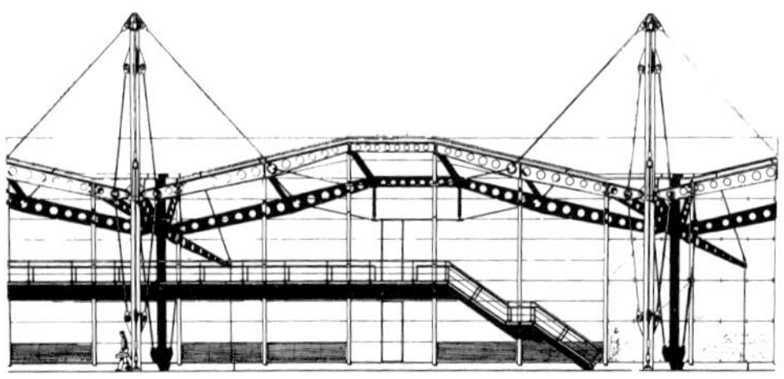

64 Norman Foster, Renault Parts Distribution Centre, Swindon, 1980–83

The building is so distinctive, Foster has claimed, that its owners have never needed to post their corporate signage upon it, though the structure is painted a vivid Renault yellow. The colour is used to delineate the primary structure, but it would hardly be unobtrusive to the colour blind.

'Pragmatic' explanations for design decisions come readily from the Foster organisation, even for the most bizarre proposals. The columns for this building are easily accounted for: 'The structure very clearly expresses the gravity forces acting on the roof'; 'the form evolved into a more accurate representation of the bending moment diagram for the beam, with the deepest section now coinciding with the mid-point support cables . . . as detail analysis was resolved, the heavier form evolved, showing fewer holes and larger joints.' This is post-rationalisation in which the post denies the rationale. In reality, the masts – more mast-like than the masts of most sailing ships – are overt references to pre-war maritime and aviation technology. They are suffused with the innocent longing for the rediscovery of boyhood enthusiasms which has for decades been the apotheosis of car design, and for this reason the architecture perfectly fits its purpose, and continues to exercise a powerful allure upon male designers of all ages.

The column bays are 24 square metres, and fifty-nine of them are planted on a square grid to cover the 30,000-square-metre facility. Each column is constructed around a 16-metre high steel cylindrical section, 30 centimetres in diameter. The four arms radiate from the mid-span of the column and each is braced by two tension rods, one very close to the column to brace the pin connection, and one at the arms' apex. Like tropical trees, the spreading boughs drop slender lianas at the outermost line of the perimeter: the building has no definitive edge. The webs of the outstretched arms are further ornamented with the popular high-tech conceit of 'weight reducing' cut-out holes. A quarter of the 14,000 tonnes of steelwork, interestingly, is accounted for by the solid pin joint connections: this is a considerable economic investment in ornamentalism.

High-tech architecture absorbed the use of the mast column from a variety of marginal sources: from Fuller's early house designs, from 'festival' architecture (particularly the iconic Skylon of the 1953 Festival of Britain), and from the fabric tension structures of Frei Otto, among others, to which mast structures are integral. It was, however, through the Renault Centre that masts were first prominently used where columns or frames would have sufficed perfectly well, simply for the ornate, gothic visual profile that they allow the architect to weave.

RICARDO BOFILL

Les Collonnes du St Christophe, Paris (1981–86)

The ancient superstition that properly proportioned columns can exert magical control over the human spirit seems, like many folkloric beliefs, to be capable of surviving in even the most stringently pragmatic of theoretical environments. In Gaudí's Park Güell, the talismanic properties of the Doric order were invoked to instil a sense of community (see pages 26–28), and it is not surprising that a similar notion should have resurfaced as a touchstone of Nazi architecture. Albert Speer and Adolf Hitler's unrealised Germania bristled with gigantic classical cylindroids, and one of their few realised civil works, the temporary Cathedral

65 The searchlight columns of Albert Speer's Cathedral of Light, Nuremberg, 1934

of Light at the Nuremberg rally (1934), was as potent as magical columnar artifice can get (65).

The same idea is revived by the Spaniard Ricardo Bofill (b1939) in his mass housing project, Les Collonnes du St Christophe. Here the classical column, of all things, is called upon to impose discipline upon the residents of the 380 family apartments that constitute this dense, social housing community (66, 67, 68).

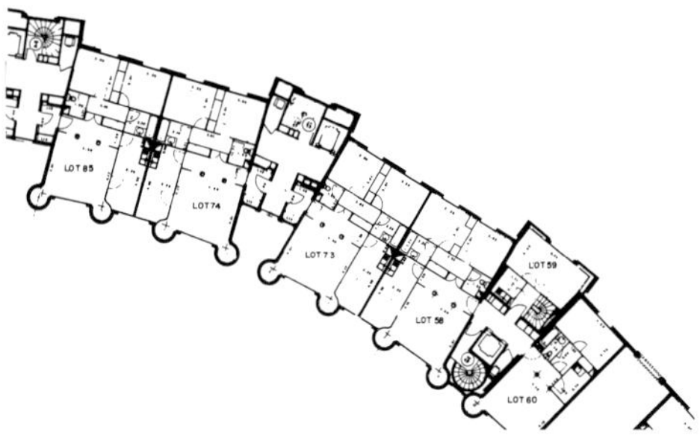

66 Ricardo Bofill, plan of Les Collonnes du St Christophe, Paris, 1981–86

Apparently, the columns are not intended to be a balm for anarchic tendencies; rather they are intended to stimulate a sense of dutiful neighbourliness among the residents. As a further entanglement of the strands of reason, the columns are not at all structural. In fact, the building has no real structural columns at all. The gigantic cylinders are hollow and, forming the external walls of the many flats, provide dutiful occupants with a useful bay-window-type space, without windows.

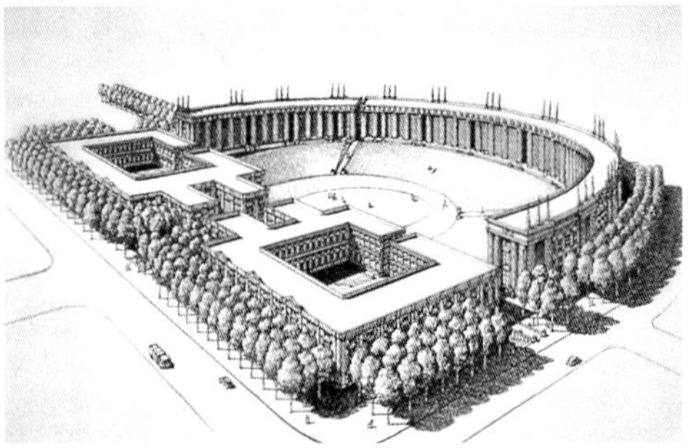

67 Perspective view of Les Collonnes du St Christophe

The columns, or more properly 'pilasters', are 15 metres high and have the overall proportions and capitals of the Doric order. However, the shafts are, like Speer's columns, parallel, without taper or entasis. The impression that they have been extruded is a consequence of the economics of construction. The facades are built up in 'slip-formed' *in-situ* concrete and from pre-cast concrete panels, all of which are identical. A taper or other inflection would, of course, require a large number of different form-works or panel shapes, and significantly greater expense. This is a social housing project, so a dead parallel profile is the only option available. This encumbers the building with a lifeless inertia that is not redeemed by economic rationale.

It is hardly necessary to observe that there are several layers of hubris in operation here. Most uncomfortably, there is the implicit belief that architects are empowered to determine social attitudes; that the overblown scale of the columns is necessary in order to impress their symbolic power on the less-than-receptive aesthetic sensibilities of the plebeian mass who inhabit the project; and that earlier, better architects would have been dogmatically opposed to this solution for any reason other than that it is transparently specious.

68 Ricardo Bofill, Les Collonnes du St Christophe

OTTO STEIDLE

The house that Charles and Ray Eames built for themselves in 1949 might have had a more immediately influential effect on architectural manners. Although it was admired, it was at first regarded as a curiosity because it had been built in the suburbs of the least cultured city in the world, Los Angeles, by a married couple who had no architectural training.

The Eames House was eventually used vigorously to exemplify the academic revival of mass-production chic in the 1960s. The house was built on a simple frame of unclothed standard steel sections. The enclosure walls were equally unpretentious standard panels, bought and fixed without modification. The whole suggested that housing, or any other building, might one day be delivered in 'flat-pack' off the back of a vehicle and easily assembled with minimum preparation.

High-tech architecture was to flower from this theoretical ideal. It is, however, rare that every embellishment of a prescriptive theory survives when the theory matures into realisation. The monumentality of the Pompidou Centre had the effect of relaxing the theoretical emphasis on low-cost mass production and functional flexibility (the capability of altering spaces, adding spatial components, moving, opening and transforming partitions and so on), neither of which it was able to realise. While the high-tech idiom developed from the priorities defined by the Pompidou Centre, a few purist designers continued to pursue the original, less glamorous functionalist dream. In London, Cedric Price (b1934) argued, with inimitable rhetoric, that building should make no investment in formal composition, as this undermines flexibility and therefore utility, and fails to address the likelihood that the functional requirements of its users are likely to change.

69 Otto Steidle, Wohnenlage Schwabing under construction, Munich, 1972–74

Wohnenlage Schwabing, Munich (1972–74)

The German Otto Steidle (b1943) has probably come closest to realising the functional idealist vision. His Wohnenlage Schwabing in Munich predates a similar, better known building made for the Kassel Documenta (1979–82). These buildings demonstrate that, however hard the designer might concentrate on objectives that are unconcerned with formality, a formal language will be the end result (69, 70).

70 Otto Steidle, Wohnenlage Schwabing

The idea behind the Wohnenlage Schwabing is that accommodation can be added and taken away as needs arise. The accommodation units, typically referred to in this genre as 'pods', are slotted into a concrete rack. The rack is never empty, and at the same time never full – if either state were reached, the building would be a failure.

The columns are precast with multiple projecting haunches. Each of the haunches is capable of supporting the end of one of the loaded cross beams, but only a very few of them will actually be loaded at any time. The unloaded protrusions give the columns something of the primal quality of rough-hewn pine trunks.

These columns are an extreme example of an archetype which is becoming more common: throughout the twentieth century, academic architects in particular have been drawn to the idea that mass production will secure the future of architecture, that it will solve all the problems of scarcity, repeated re-invention, and antiquated technique that tie the profession to the past. But in so far as architects like Otto Steidle have played formal games on the basis of this theory, admittedly with quite pleasing results, their efforts are tangential to the real consequences of the industrialisation of building.

The transition that has been invited for so long is gradually taking place. System building inexorably accounts for greater and greater quantities of built work and is slowly, depressingly eroding the richness of our built environment. None of it is glamorous, quirky or engaging like the Wohnenlage Schwabing: little if any of it is being designed by architects.

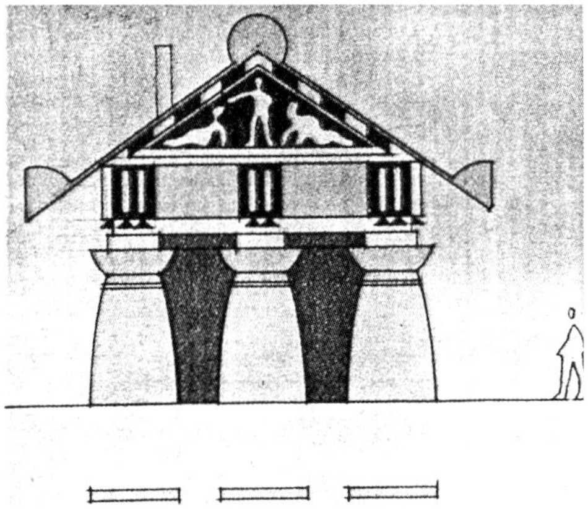

71 Robert Venturi, Eclectic House project, 1975: elevation and plan of portico

ROBERT VENTURI AND CHARLES JENCKS

Post-Modern Neo-Classicism (1975–83)

Venturi's indispensable thesis, *Complexity and Contradiction in Architecture (1966)*, was written as a counterpoint to the assertive philistinism of progressive architecture in the 1960s. After decades of dogmatic denial, Venturi (b1925) made the radical suggestion that old buildings were compositionally more sophisticated and might in the future provide the elements of an enriched modern vocabulary. His paper languished on the margins of American academia, until eventually in the mid-1970s the profession caught up with him; within a few years every city in the Western world had been enhanced with plywood pediments and stick-on 'classical' stumps.

In recent years, the discipline of evolutionary psychology has overturned conventional understanding of human behaviour. It explains instinctive actions in terms of evolutionary advantage: what it cannot do is make any suggestions as to how we can change these behaviours. Venturi's thesis is at the centre of a similar crisis: it is a wonderful example of the difficulties that arise when an observational essay is transformed into a prescriptive theory of design.

72 Doric columns in the courtyard of James Stirling's Staatsgalerie, Stuttgart, 1980–83

The column paradigm that emerged from the Post-Modern Neo-Classical idiom is of no particular form or structural purpose; it is distinguished by its superficial references to the classical orders, and usually by a deliberate lack of both physical and conceptual solidity. The ultimate Post-Modern Neo-Classical column is simply a two-dimensional cut-out. Venturi's own Eclectic House project of 1975 gave credence to this formal conceit: its cheapness led it to become the most popular store-front decoration in the Western world, for the few years it took for the plywood to warp (71).

The architectural profession quickly tired of the 'wit' of Post-Modernism. All jokes die after a few re-tellings. The style was highly fashionable for a few years and was assimilated more rapidly throughout the design disciplines than any earlier architectural style, a fact that is testament to its simple commercial appeal.

Post-Modernism produced no masterpieces that stand comparison with the great works of earlier twentieth-century movements. The Post-Modern Neo-Classical sub-genre, identified by its highly visible proponent, the critic Charles Jencks (b1939), has yielded one or two memorable exemplars, the most familiar being

James Stirling's Staatsgalerie in Stuttgart of 1980–83 (72). Stirling's pot-bellied Doric figures are moulded in reconstituted stone and are, at least, three dimensional, although according to the rules of the genre, their solidity is no great virtue.

In his 1981 eulogy *Post-Modern Neo-Classicism* Jencks draws particular attention to Venturi's buildings and those of fellow Americans Michael Graves, Charles Moore and Robert Stern, highlighting the American origin of the idiom. In reality, James Stirling, Aldo Rossi, Quinlan Terry, Leon Krier and other Europeans with whom the Americans have been grouped, all have their own eccentric historicist standpoints and their merging is opportunistic. Terry, for example, is an extreme eighteenth-century revivalist who has been known to endorse John Wood the Elder's belief in the heavenly origin of the orders, whereas Rossi comes from the rationalist tradition, and Krier, according to some, from the quasi-fascist (73, 74).

In the mid-1980s, the neo-classical revival received the unprecedented (and perhaps ill-advised) endorsement of Prince Charles. One of the consequences of this was the re-allocation of the London National Gallery extension project from the legitimate winners of the design competition, to Robert Venturi himself. This, the academic's largest and least transient commission, was to result in massive anticlimax: with the completion of the project the highest expectations of the Post-Modern Neo-Classical idiom eventually gave way to acceptance of its morbidity.

In the cycles of classical revival and decline, it is, paradoxically, the purist functionalism of the high Modern Movement which, in our century, represents the real revival of classicism. The proof of this is in its emphasis on social morality. Even though it claims to be a classical revival, it is the explosion of Post-Modernism, with its emphasis on gratification, that prefigures the inevitable slide into decadence and fragmentation. This is where we find the profession today; without direction, awaiting the *fin de siècle,* and the revival to come.

73 Aldo Rossi, Hotel Il Palazzo, Fukuoka, Japan, 1988

74 Opening Day, visualisation of Leon Krier's Atlantis Holiday Resort, Tenerife

CHAPTER 4

MONOLITHS

75 Grand Menhir Brisé, Locmariquer, Brittany, *c*2000BC

For several centuries either side of 2500BC, some complex and ambitious building projects were carried out around what is now north-west Europe. The commonplace construction work of a stone-age community would involve the use of large amounts of green timber, dung and straw, and the shifting of mountains of earth. The special constructions, the *grands projets* of the day, were made of hewn, erected stone. The monolith Men-er Hroec'h, or the Grand Menhir Brisé (*c*2000BC), at Locmariquer in Brittany was almost certainly the heaviest object ever moved on the earth's surface without the aid of machinery (75).

The stone was erected on a peninsular, the sea on three sides, where it probably acted as a 'fore-sight', aligned with other smaller stones inland and the position of the moon-rise on calendrical high days. The 22.5-metre-high, 330-tonne monolith was carried over 50 miles to the peninsula from the site at which it was quarried, and once located, it was sunk 3 metres into the ground. It is obvious that considerable care was taken in positioning and consolidating the monolith, which makes it all the more surprising that the enormous stone now lies broken in four pieces. Calculations have established that the force required virtually to throw the monolith from its position into that in which the pieces lie could only have been provided by a violent earthquake, an exceptionally rare event in this tectonically stable area.

The felling of a great monolith is an apocalyptic portent. The aspirations of a civilisation, and a considerable part of its manpower, are invested in such structures and the destruction of the column is inextricably associated with great death, or defeat; it is as if the collective body of the civilisation is felled, and the world will never be the same. It is for this reason that the stones of the ancient world, the monoliths, obelisks and colossal statuary, as well as being spectacular, are so profoundly moving. Percy Bysshe Shelley's romantic sonnet 'Ozymandias' evoked this pathos in his Victorian readers:

I met a traveller from an antique land
Who said: two vast and trunkless legs of stone
Stand in the desert. Near them, on the sand,
Half sunk, a shattered visage lies, whose frown,

And wrinkled lip, and sneer of cold command,
Tell that its sculptor well those passions read
Which yet survive, stamped on these lifeless things,
The hand that mocked them and the heart that fed:
And on the pedestal these words appear:
'My name is Ozymandias, king of kings:
Look on my works, ye Mighty, and despair!'
Nothing beside remains. Round the decay
Of that colossal wreck, boundless and bare
The lone and level sands stretch far away.

The builders of the Megalithic period (3500–1500BC) typically enhanced the utility of their monoliths as astronomical markers with gender symbolism, for example in the alternating slender and broad stones that are found in some alignments, as well as with more overt carved pictograms, hollows and piercings. This they have in common with the classical orders and with every other column: their meanings remain consistent throughout subsequent millennia. If there was any doubt that monoliths, whatever their purpose, are all primarily representations of idealised living humans, colossal statuary is unambiguous. While there is obviously a convergence with domestic sculpture, colossi tend to be distinguished by stylised abstraction of the figure, often to the extent that it is confined within a cylindrical or rectangular volume, to emphasise the columnar nature of the monument.

The colossi (or *moai*) of Easter Island (*c*AD1000–1500) became so much the focus of the identity of their builders that every one of the island's trees was felled for use in transportation, and the island had to be abandoned to its stone legion (77). For the few colossi and other monoliths that remain in Mexico (76), Egypt and elsewhere from ancient times, many more have been felled, broken or purloined, usually by conquering armies, in the most emphatic gesture of subjugation.

The association of living body with monolith is also the origin of the custom of marking the grave of a human being with a standing stone (78). The stone is the incorporation of the spirit of the departed; as long as it stands, the presence of the deceased remains among the living and it is for this reason that the destruction of

76 Toltec colossus, Mexico, 13th century

77 Colossus, Easter Island, *c*AD1000–1500

headstones is still regarded as the most barbaric affront, far worse than the mere destruction of property.

The technicalities of basic signal monoliths are, in general, unsurprising. The great majority that survive from ancient times are wrought from solid stone or, like the totem poles of the American Indians, carved in timber, with the simple engineering specification that they should be founded upon a stable base. The more expository *details* are those that convey the talismanic value of the artefacts through the efforts that have been made to build, to capture and to display them, as 'living' witnesses to the tides of human civilisation.

For the further purposes of taxonomy (and without suggesting that these two categories are in any way definitive or clearly delineated), the monoliths may be divided into two families. Even the oldest known monoliths had a purpose: they had a role in mapping the night sky and, presumably, great religious significance. The obelisks of ancient Egypt were usually erected in pairs and acted as talismanic 'guardians' to the entrances of temples. The monoliths of this first family convey their 'message' by the basic means of their presence, their shape and through surface ornamentation if any is present. This family could be called the Basic Signal Monoliths: recent manifestations of them are rare.

The second family comprises a variety of structures that share the attributes of columns but also engage their environment in a direct way. Probably the earliest, and still the definitive example, is the lighthouse. Although obviously a 'monolithic' form, the column broadcasts its message by throwing out a light from its capital, a light which might be useful to navigators, but which is also, as one would expect of such an artefact, powerfully symbolic. The family of Complex Signal Monoliths includes any vertical structure that has no contents, but which exists to elevate something active to the position of the capital, for the purposes of amplifying its message, or to interact in some way with the natural environment which would not be possible down amongst earth-bound mortals.

78 Hybrid figurative Christian monolithic headstone, Llantwit Major, Wales, 6th century

SOME BASIC SIGNAL MONOLITHS

The purpose of Columns is to rise above mortals.

Pliny the Elder (*c*AD70)

The three definitive components of the column, its base, shaft and capital, are not present in the megalithic stones. However, these features do manifest themselves, independently of Hellenic precedent and often before it, in figurative and monolithic columns of prehistoric cultures in the Americas, the Pacific Islands, North Africa and Mediterranean Europe.

In the fourth century BC, when Alexander the Great extended his empire into north Africa, he did not single-handedly bring an end to two-and-a-half millennia of Egyptian pharoanic tradition. Before his conquest, Egypt was briefly a Persian province. It was, however, Alexander's influence, and that of the Romans who took control a mere three centuries after him, that finally erased the surviving remnants of Egyptian cultural tradition. Only the vast monuments, with as yet secret treasure hoards, remained.

While the Greeks did not make a habit of draining conquered territories of their riches, the Roman invaders who followed had more voracious manners. The port of Alexandria, the base of the Greek Ptolemaic pharaohs, remained the centre under Roman rule and slowly most treasures that could be moved were brought there from the dying cities that lay along the Nile, and thence from Alexandria to Rome.

A few Egyptian obelisks and Roman replicas can be found scattered across the Roman Mediterranean. The majority ended up in Rome itself and, apart from the large obelisk transferred to Constantinople and one small stele that was taken to ornament the Pitti Palace in Florence during the Renaissance, all the obelisks that are known to have been taken there still remain in the city. The oldest and, at 31 metres high (including its sixteenth-century base), by far the largest of the monoliths stands in the Piazza San Giovanni. It dates from *c*AD1500 and was brought to Rome by Constantine II in AD337. Most of the other stones are associated with the reign of Augustus, in the first century.

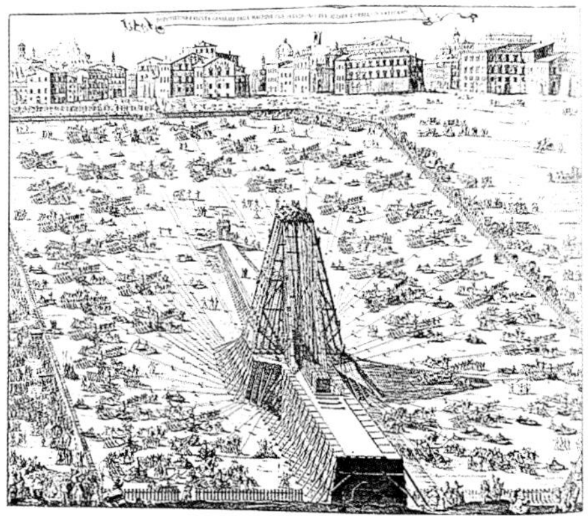

79 Woodcut by Carlo Fontana showing the erection of the Vatican Obelisk in St Peter's Square, Rome, 1586

The San Giovanni obelisk, the Roman replica at Piazza dell'Esquilino, and the large obelisk that stands in the centre of St Peter's Square have a more recent association with Pope Sixtus V (1585–90), the first Counter-Reformation pope. It was Sixtus V who assumed the job of restoring the city from the shattered state in which it had been left fifty years earlier by the marauding army of Charles V, and at the behest of his architect, the extraordinary Domenico Fontana (1543–1607), much of the restoration was focused on the re-erection of the Egyptian obelisks, many of which had fallen centuries before Charles V sacked the city (79, 80).

The obelisk in St Peter's Square is unusual. Although it is cut from pink granite like most obelisks, it has no inscriptions. At 27 metres high it is also larger than most. Without hieroglyphs its provenance is uncertain, although it is thought to have been brought from Heliopolis to Rome by Caligula around AD37. The new basilica of St Peter (1506–1626), at the centre of the Vatican, was the most important building in the Christian world. Perhaps the anonymity of the stone added, somehow, to the aura of its ancient spiritual power. Whatever the reason, in 1586 a massive effort was devoted to the removal of the fallen stone from the Roman basilica and to its re-erection in the geometrical centre of the new basilica's piazza.

80 Vatican Obelisk, Rome

Without the ability to test proposals by calculation, the method eventually employed, at vast expense, required 900 men and 115 horses to turn 48 winches, a slipway and a towering armature threaded with the cables that transferred the work of the winch teams to the load. The location and the improvised ritual of the raising of the stone attained such sacred significance that Sixtus V is alleged to have forbidden any of the labourers to speak during the erection, on pain of death. Local folklore maintains that one heroic participant did shout out, for water to drench a cable which was about to yield. Instead of executing him, Sixtus rewarded him for saving the project from disaster. The logistical particulars were recorded in a series of woodcuts a century later by the architect's descendant, Carlo Fontana (79). If the scale of the effort appears somewhat excessive for the task, it is because efficiency was not, at the time, an element of engineers' thinking. At the end of the sixteenth century mechanical design was, like architecture, still concerned primarily with aesthetics and the revelation of the microcosmos. The raising of the Vatican obelisk is a throwback to megalithic times, the rite of the execution confers significance on the monolith, the scale of the effort befits the spiritual enormity of the city's endeavour.

81 Trajan's Column, Rome, AD113

The Vatican obelisk was one of several that were either 'rediscovered' or restored from great supine fragments during the sixteenth and seventeenth centuries. Among these were a few extraordinary native columns, the landmarks of Trajan (*c*AD113) and Marcus Aurelius (AD174), and the smaller column of Antonius Pius (second century AD), which is a true monolith, carved from a single stone (81, 82).

The Romans were the great assimilators. Their native invention is no less remarkable than that of the cultures they absorbed, but the clear impression of the relative status of the most popular Roman building types is that the Greek or, latterly, the Egyptian forms had the greater integrity. Although they are usually classified as Tuscan order, the triumphal columns of Trajan and Marcus Aurelius are a Roman re-invention of a Hellenic form. Each is a freestanding cylinder with authentic entasis, built up from twenty-eight marble drums around a central spiral stair. Both columns are approximately 30 metres high (100 Roman feet) and average 4 metres in diameter. The surface of each is covered with a helical relief depicting victorious campaigns. Trajan's is the more finely sculpted; the montage of his exploits in Dacia (Romania) is also pierced with small openings which allow light into the internal stair.

82 Marcus Aurelius' Column, Rome, AD174

Both columns were threatened by decrepitude until 1589 when, again by the intervention of Pope Sixtus V and his architect, Domenico Fontana, they were Christianised for their own preservation. Each had its imperial statue replaced by a Christian saint. Indeed, to demonstrate to the city that Trajan's Column's miraculous properties were not diminished by the change in religion, it became the centre of a remarkable rescue when Pope Gregory the Great (Sixtus' successor) was sufficiently moved by carvings depicting the Emperor's compassion that he offered prayers for Trajan's soul. Soon after the Pope experienced a vision confirming that Trajan would indeed be released from damnation, on the condition that the Pope should not pray for the release of any more pagans from their torment.

A few decades after the obelisk revival of anti-Reformation Rome, the ardently anti-papist citizenry of London were struggling to restore their own city from the Great Fire of 1666. Although few lives had been lost, a major part of the old city was razed to the ground and the task of repair was, to say the least, daunting. The King and his government pondered the best means of making good the damage and decided that the appropriate course of action would be to build a monumental column (83).

83 Engraving of Sir Christopher Wren's Monument to the Great Fire of London, 1671–77

The column (1671–77), designed by Sir Christopher Wren's Office of Works, was intended as a statement of the city's resurrection, the stricken body dragging itself back to its feet (despite the worst efforts of the papists whom the column's inscriptions blame for the disaster). This symbolism is given a precise expression by the decision to locate the Monument 60.2 metres away from the spot at which the fire was believed to have started, and to build it 60.2 metres high.

London was faced with the prospect of having to replace over thirteen thousand houses, eighty-nine churches, and dozens of amenities and civic buildings. Notwithstanding this, after barely four years of rebuilding, a Royal decree granted Wren's towering Doric column the first choice of all stone produced by the Isle of Portland until it was complete. Because, like the city's new churches, it was financed by a levy on coal traded in the Port of London, for six years until its completion in 1677, the construction of the column also further inflated the price of the city's fuel.

Finance from the coal tax came slowly: the budgetary constraints prevented the commissioning of a better artist than Caius Gabriel Cibber to carve the ornament at the base, and his allegorical reliefs, like those upon Marcus Aurelius' Column, are charmingly second rate. The column itself is, however, superbly modelled, and the crown, which but for lack of funds would have had a statue of the king, instead has a unique vase of flames, a symbol invented by Wren, and, as his correspondence suggests, discreetly preferred for its relevance to the event it commemorated. It was, after all, the city that was to be resurrected and immortalised, and not, on this occasion, Charles II.

Until the building of the Washington Monument, the Monument to the Fire of London was the highest column in the world, substantially taller than the more visible 44-metre Nelson memorial column (1839–42) designed by William Railton. It is still the largest in a classical Hellenistic form.

In 1798, two centuries after the raising of the Vatican Obelisk, Napoleon invaded Egypt, bringing to an end the three-hundred-year Turkish Ottoman domination of the ancient kingdom. Barely one year later, Napoleon was, in turn, evicted from the country by the British Navy under Nelson, and Egypt entered a period of leaderless

84 The gateway of Ramses II, Luxor, Egypt, *c*1300BC, with its one remaining obelisk

85 The barge built to transport Cleopatra's Needle from Alexandria to England in 1878

86 The erection of Cleopatra's Needle on the Thames Embankment, London, 1878

turmoil. Control of sorts was eventually restored by a contingent of the Turkish army, under a general, Mohammed Ali, who declared himself Pasha in 1805.

The European powers exerted a compelling influence upon his new country's fortunes. Cut off from the declining Ottoman powerbase, the Pasha responded to the lack of native resources by beginning foreign campaigns of his own, and simultaneously inaugurating a modernisation drive. He rapidly accumulated massive debts. His borrowing from Britain and France left him still more exposed to foreign interference in Egypt's affairs.

Without the means to repay the loans, the Pasha looked for suitable symbolic tokens with which to retain the favour of his creditors. The submission of some of the few remaining great obelisks cost the Pasha nothing, but conveyed perfectly the necessary frank admission of obeisance. So began the second and final wave of Egyptian obelisk dispersal; the long journeys of the two stelae of Thothmes III, one to New York, and one to London, and of Ramses II's splendid gateway stele from the temple of Luxor to its new home in Paris (84).

The removal of the two neglected stones from Alexandria was unfortunate; they had originally been erected for Thothmes III at Heliopolis around 1500BC, but the Romans had looted them to the new port city in 10BC, and their further removal did no more than compound Roman negligence. However, notwithstanding the fact that they were given the pair of obelisks and took only the one, the removal of the Luxor stele by the French, after three thousand years of standing unmolested, appears now to have been an appalling desecration.

The French and British gifts were announced in 1819. The Pasha quite reasonably suggested that the recipients should come and collect their prizes. The British considered the cost and left their obelisk where it lay. It was only after the new French regime of Louis Philippe, the citizen king, removed and re-erected theirs – the largest obelisk – in the centre of the Place de la Concorde with extravagant ceremony that the British were provoked into action, on the initiative of General Sir James Alexander. Even then, decades were to pass before the English project found a backer, and it was a surgeon, Erasmus Wilson, who made the major contribution, reflecting the great fashionable interest in Egyptology that pervaded English society.

In 1878, after protracted planning and preparation, the Alexandria-based engineer John Dixon succeeded in manoeuvring the remaining obelisk into an extraordinary cylindrical wrought-iron barge, which was intended to hold safely the 20-metre, 186-tonne load for a sea journey direct to England (85). The exporters drastically underestimated the logistical difficulty of their project. The ocean-going barge broke free from the towing ship in a storm in the Bay of Biscay, causing the loss of six seamen; the stone was also lost for a while before being recovered and taken into a Spanish port. A second steamship was dispatched from London to escort it, and the convoy was greeted by cheering crowds when it eventually reached the Port of London. The ingenious machine used to erect the obelisk attracted as much interest as the stone itself (86). A 'time-capsule' containing other modern marvels such as a railways guide and a photograph album of society beauties was interred below the stone. In Victorian terms this was, no doubt, an appropriate embellishment.

The monolith is enhanced by these indignities: it is an unintended memorial to the hubris of Victorian society. Even so, the present siting of Cleopatra's Needle, overshadowed by a railway bridge and isolated by traffic against the wall of the Thames embankment, evokes a sense of pity that one would normally feel for a fellow human.

America set out to acquire its own obelisk in 1879. Although it met with popular support, the venture was instigated by the community of New York Freemasons, who could not tolerate the condescension of Europeans. The American diplomatic presence in Alexandria took a year to browbeat Tawfiq Pasha into giving up the second Alexandrian stele. The European community of Alexandria sensed the true motives of the American endeavour and tormented the expeditionary party almost to the outbreak of hostilities before it grudgingly allowed the Americans to leave with their trophy. The obelisk was set up in Central Park in 1881 using the innovative British technique (although this was not credited at the time). The Egyptian State, still cash-starved, was to enjoy the benefit of American gratitude: the magic of the stones was still potent.

Gustave Eiffel (1832–1923) re-wrote the conventions of monument building with his 300-metre iron-frame column of 1889 (87). His project was at the crossroads of architecture: it showed that technology was not just going to change the forms of functional buildings, as the Crystal Palace had demonstrated, but that it also provided the means to make monumental forms of fearsome height and discorporate, skeletal countenance which, far from expressing the collective will of humanity, expressed its insignificance.

The awesome achievements of skyscraper builders in Chicago and New York also helped to bring an end to the civilised custom of building simple signal monoliths. Whether tall signal structures were proposed for advertising products, as in the USA, or politics, as in Alexandra Exter and Boris Gladkov's 1923 Constructivist monument at the Russian Agricultural Exhibition in Moscow (88), it was the technology of construction that was celebrated, and the wonder of all things synthetic.

The axis powers further tainted the custom of monument building by reviving it wholeheartedly and with appalling vulgarity. Mussolini even looted a fourth-century Ethiopian sacred column, the Obelisk of Axum, and brought it to Rome where it still stands, exuding an aura of sadness.

87 Gustave Eiffel, Eiffel Tower, Paris, 1889

88 Alexandra Exter and Boris Gladkov, Constructivist monument, Moscow, 1923

89 Luis Barragán, Satellite City Towers, Mexico City, 1953

90 Anthony Gormley, *Angel of The North*, 1995–98

91 Watercolour by Harold Oakley showing a reconstruction of the Pharos Lighthouse, completed 280BC

In the latter half of the twentieth century, it is very difficult to think of simple signal monoliths as ever being able to avoid the faint stench of their recent association with the Nazis, particularly if they are of a neo-classical form, and proposals of this kind, when they occasionally surface are treated with justifiable derision. If the call goes out, these days, to find a design for a monument, the emphasis is always upon the appropriateness of the form, a constraint which is generally interpreted as excluding simple signal monoliths, but allowing big wheels, unnecessary new bridges and the like.

There are very few modern simple signal monoliths that avoid distasteful associations. One that does is Luis Barragán's stunning quintet of monoliths, the Satellite City Towers (89). These were built on a freeway island on the outskirts of Mexico City in 1953. They are polyhedral forms, triangular in plan, cast hollow using simple slip-formed concrete. Where Eiffel's monument, and so many later technological triumphs seem to diminish humanity and contribute to the sense of alienation that pervades contemporary society, Barragán's monument seems actually to be sympathetic to the metropolitan predicament and to have at least some of the totemic qualities of its ancient precursors.

The late twentieth century also accepts unquestioningly that non-figurative forms can be sculpture, and it is perhaps the case that if a large, functionless edifice is to be built it is, these days, more properly a job for an artist to carry out (with the obliging assistance of engineers). Indeed, the artist Anthony Gormley's colossal *Angel of The North*, recently erected in Gateshead, would appear to be just such an enterprise. This is a monumental standing figure with vast outstretched wings, all in steel; where the Mayans, Egyptians and many other ancient peoples confined their colossi to simple geometrical volumes, because those basic forms had their own mystical power, Gormley appears to have forged his figure within the modern sacred outlines of an aeroplane (90).

SOME COMPLEX SIGNAL MONOLITHS

Nearly a thousand years after the high point of the New Kingdom period of Egyptian civilisation (the origin of the great hypostyle of Karnak), the native Egyptian dynasties came to an end, overwhelmed first by Persia and then, around 300BC, by the Grecian army of Alexander the Great. The pharaoh Ptolemy was Greek; he had been one of Alexander's generals, but according to Egyptian convention, he was deified. The Ptolemaic Dynasty, which he founded, was based at Alexandria, the Nile port named after the immortal Alexander. It was to Ptolemy II that the first known lighthouse was dedicated upon its completion in 280BC. This 130-metre high monolithic column had, as its capital, a permanently illuminated beacon.

(Current thinking suggests that the beacon was lit at ground level and reflected from the top of the structure with a metal mirror.) Before the Pharos was built, beacons would customarily have been lit at the top of solid shafts of timber or stone, and the analogy to the capital would have been much less ambiguous. The Pharos has a hollow masonry shaft and a temple in its base, but its light is, like the capital of any column, the part that is closest to the higher realm: it is the point of the interface of the earthly with the sacred (91).

The fact that the light was also a guide and a guardian would only have enhanced its significance, and the evidence that remains of the long demolished structure suggests that it was protected by the kind of colossal statuary and inscription that was characteristic of high Pharaonic temples: it was no simple flood light. The Pharos, as well as being the first lighthouse, is likely to have been the first monolith in which a sacred, ceremonial prominence was augmented by an altruistic earthly purpose.

It is not just superstitious mariners who recognise talismanic qualities in the elevated beacon. It is known that lighthouse builders in ancient Libya, further along the north African coast, whilst never working on the scale of the Pharos, always respected the sanctity of the beacon by building temples below their columns, and to this day we think of lighthouses as having an almost spiritual protective aura and as embodying our collective humanity: qualities which make explicit their monolithic ancestry.

Most lighthouses are built on shore, but it is the offshore structures that have the more exacting structural requirements, for example, the need to withstand wave pressure of 20 tonnes per square metre. By being designed to meet these loads they have become more purely monolithic. This cylindrical archetype originates with John Smeaton's Eddystone Rock beacon, Plymouth, of 1756–59 (92, 93, 94).

To make this column, the rock on which it stands (which is barely 10 metres long) had to be cut to receive granite blocks. The blocks, which weighed up to 3 tonnes, were taken out on small boats to the tiny island to be dovetailed into the natural rock, and locked in place with toggles and plugs. Although three years elapsed between commencement and completion, the climate and tides meant only sixteen weeks work was possible on site. Taking this into account, and the fact that there was barely room for a dozen men on the rock at any time, Smeaton's achievement almost defies belief. It is interesting to compare the efforts of Smeaton's small team with those of Domenico Fontana's silent legion. Pulley blocks had been a familiar tool for centuries, but at some point between the sixteenth and eighteenth centuries their usefulness came to outweigh aesthetic considerations.

Smeaton's column was not the first to have been built on the rock. Two previous structures had been short-lived, demolished by high seas. Smeaton's design was by far the most enduring. It provided over 120 years of service before it had to be

92 John Smeaton, Eddystone Rock beacon, Plymouth, 1756–59

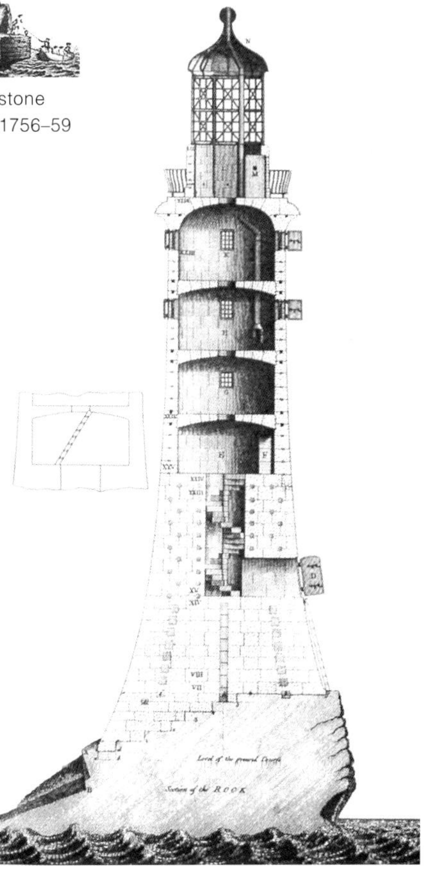

93 Section through Eddystone beacon

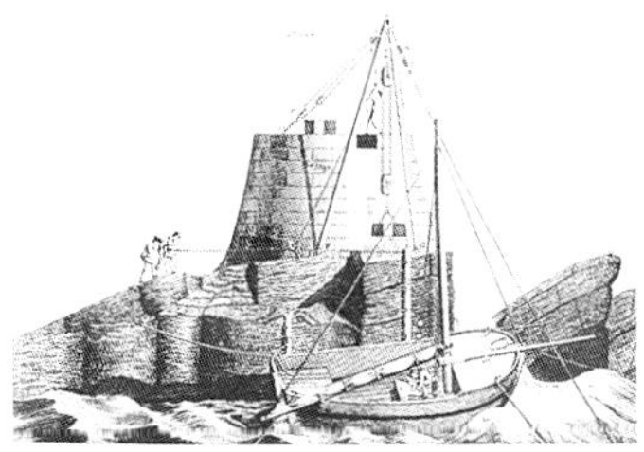

94 Construction of Eddystone beacon

95 Ronald Ward and Partners, Dungeness Lighthouse, Kent, 1960

96 Minarets of the Mosque of Muhammad Ali, Cairo, 1840

97 Cooling tower, Schmehausen Nuclear Power Station, Germany

dismantled. Even then it was the disintegration of the base rock that caused the instability. The column was as strong as the day it was built.

The cross section of Smeaton's column (93) shows with uncomfortable clarity the dungeon-like living spaces that were home to the light's keepers. The keepers were imprisoned in their cells, utterly isolated from land – often, in the hardest season, for weeks on end. It should not have been a coveted job, and yet they were universally admired for their altruism and their real heroism. In many respects they embodied the priestly, abstinent tradition of their ancient predecessors.

The majority of lighthouses have always been onshore and their crews not imprisoned as they were on offshore beacons. The last offshore lighthouse went automatic only within the last decade, long after it would have been possible to automate.

Today, around British coasts at least, manned lighthouses are a thing of the past, and the columns have regained their lonely monumentality. The last lighthouse to be built on the British coast was constructed in 1960 at Dungeness, Kent, and it was, surprisingly, designed by an architect, Ronald Ward and Partners (95). It is the architect's involvement, purists have often complained, that explains the spiral ramp at the base, which serves no purpose and is merely ornamental. This is correct, but the base does befit the column, and no columns, not even lighthouses, are primarily functional.

An elevated beacon is a common feature of a complex signal monolith. The beacon is the capital of the column. The steeples of Christian churches are also capital-like features; like the light from a beacon, the peal of bells from the top of the monolith has the practical purpose of communication. It broadcasts the call to prayer and, at the same time, conveys the symbolic suggestion that the sound of the bells can be understood, in a way, to emanate from the higher realm. This is also true of the minarets of Islamic mosques, from which the voice of the muezzin is projected (96).

The elements of monolithic columns are unexpectedly evident in a number of contemporary vertical structures. The wind turbines that are built more and more frequently on open hillsides are stirringly impressive in ways that electricity generators generally are not. The smoke stacks and cooling towers of industrial plants are similarly imposing (97). Le Corbusier recreated a cooling tower at Chandigarh, so profoundly inspired was he by a cluster of conventional parabolic shafts. The capitals of cooling towers are the dense clouds that emanate from their open mouths: the suggestion of a connection with the heavens could hardly be more clearly expressed.

In the Industrial Revolution, factory owners transformed the landscape of the country with hundreds of freestanding masonry-built smoke stacks (100). Very often these were constructed with classical ornamentation on the shaft and at the top, with elegant entasis or simple taper for structural stability, and with a plinth-like furnace at the base. The industrial landscape of the mid-nineteenth century bristles with monoliths and, as at San Gimignano, competition between adjacent factory owners was expressed in the race to build taller and more elegant chimney columns. Sir Giles Gilbert Scott (1880–1960), an anti-Modernist and a committed believer in overt functional classicism, believed that a Doric form was appropriate for the very largest chimneys. His original proposal for the landmark Battersea Power Station in London (1932–34) was a square Doric (98). This gave way to the present, delicately poised, fluted cylinders, now a cherished but long-redundant feature of the London cityscape.

Progressive technology has been a fruitful source of complex signal monoliths. There are strong column characteristics in telecommunications structures, with their elevated transmitters or receiving dishes, and in the revolving radar beacons that make prominent landmarks at most airports. The most noble example, and arguably the ultimate twentieth-century monolith, is the Saturn V rocket (99). With the tiny sacrificial bodies of its crew heart-breakingly crammed into the capital, it is a column that both implies the infinite perpendicular to the heavens, and fulfils our deepest spiritual yearning by tearing free of earth to traverse it.

Today the world has more complex signal monolithic column archetypes and hybrids than ever before, and it is futile to attempt to catalogue them here. The following case studies do, however, give appropriate space to some of the more engaging contemporary types, to illustrate techniques of construction that are, as a rule, outside of the realm of architecture, waiting to be brought in.

98 Sir Giles Gilbert Scott, Battersea Power Station, London, 1932–34

99 Apollo II, a Saturn V rocket

100 The factory landscape of the Industrial Revolution, chimney stacks in Wolverhampton, South Staffordshire, *c*1866

CONCLUSION

Oh, what a wonderful day, when the wall parted and the column was born.

Louis Kahn

Entasis, the thickening of the shaft of the column, is usually considered an optical correction in response to the eye's requirements for beauty of proportions. (It) also has a practical value: it increases the column's bearing capacity and its resistance to bending moments. In good architecture, structure and aesthetics meet in a natural way.

Sandaker and Eggen, *The Structural Basis of Architecture* (1993)

The two quotations above are chosen to exemplify the polarities of architectural thinking. The dogma represented by the second is an invention of the twentieth century. Nonetheless, it pervades architectural thinking to the extent that a statement such as 'In good architecture, structure and aesthetics meet in a natural way' is presented as if it has been a law of architecture since mankind started to build.

Our usefulness as designers depends on how we respond to mythologies. We may accept the 'neo-classical' myth; it is at least the genesis of the practice of 'Architecture'. We may be attracted to the minimalist myth, because it seems more rational and refined; or perhaps the allure of high-tech will be too hard to resist. However we find ourselves designing, it is essential that we appreciate that no mythology is right, none is moral, none is 'good'.

Good architecture is *not* the natural merging of structure and aesthetics. This may be what engineers would prefer it to be, because, if it were so, it would enshrine that discipline at the centre of the creative act: a privilege to which engineering is not entitled.

Look again at the second quotation, from a well-known textbook on structural design: it invokes entasis as an example of how the ancients allegedly designed their structures economically, so that the thickening of the column's shaft would resist the greater bending moment in the mid-span of the column. The presence of entasis is, according to these authors, empirical proof that classical architecture really is 'good'. However, the columns of which they speak are simple stacks of marble blocks. They have no bending action; they are balanced without foundation;

in fact, they are inclined to fall over rather easily. To reiterate, we take from the mythology that which we wish to find: there are no absolutes, no truths, no reliable formulae. There is only human imagination, tradition and the hunger we all share for fresh experience and original insight.

With the first column came the idea that the creative act of building could obviously, majestically transcend utility; that buildings open to the sky could be life-enhancing (more so than any shelter), and furthermore, that they could express the hopes and the dreams of civilisations. There are no rules and no formulae: but when designing columns it would do no harm to bear their primal *purposelessness* in mind.

What a wonderful day, when the walls parted and the column was born.

101 Gustave Doré, *Samson Collapsing the Temple*

1 William Frith, *The Railway Station*, 1863 (reproduced courtesy of Leicester City Museums)

MATTHEW DIGBY WYATT (WITH ISAMBARD KINGDOM BRUNEL)

PADDINGTON STATION

London (1851–54)

One of Brunel's colleagues on the Royal Commission for the Great Exhibition was the architect Matthew Digby Wyatt (1820–77). Prior to his involvement in the exhibition, Wyatt had published several radical papers promoting the advantages that might accrue if engineers and architects were prepared to share their skills. It was this proposition that brought him to the attention of the exhibition organisers, and into contact with Brunel. They collaborated on the shed for Paddington Station. *The Builder* magazine reported the collaboration:

> The principle on which they set out was to avoid any recurrence of existing styles, and to try the experiment of designing everything in accordance with the structural purpose or nature of the materials employed – iron and cement.

The building that Brunel and Wyatt made is remarkable, ample proof of the potential of the convergence of disciplines that Wyatt had advocated. More remarkable still, however, is the impact that the innovation of collaboration was to have upon the architectural profession. While the imagination of Brunel and Wyatt has rarely been equalled, from that moment on, 'structural purpose' became a limiting aesthetic force. The priorities of column design were inverted, the artistic prerogative subverted by 'adequacy'.

The technology for cast-iron columns and frames had been refined by specialist glasshouse manufacturers to service the demands of the aristocracy during the early nineteenth century. Because architects tended not to be involved, the columns of glasshouses – the Crystal Palace included – tended to be basic cast-iron cylinders with minimal ornament in their brackets. Wyatt had supervised the construction of the Great Exhibition house, but the concept design was a joint venture between the Fox Henderson foundry and the glasshouse builder Thomas Paxton.

Glasshouse building was not considered to be architecture. However, while architects were generally suspicious – as they always are – of the new technology, cast iron did have some influential advocates in the first decades of the nineteenth century. The maverick John Nash (1752–1835) was the best-known architect who had been prepared to use it in the construction of columns, albeit as a stone surrogate. The columns of his Carlton Terrace, London (1826), are famously cast iron, but anyone unaware of the fact would very likely assume that they are masonry and stucco.

By contrast, Wyatt had been emphatic in his pamphlets that cast-iron columns were incompatible with the classical orders. When Brunel invited him to collaborate on Paddington Station, it was because he hoped that Wyatt would deliver the new order that the architecture of the Industrial Age required. Although they did have the effect of giving momentum to the embryonic culture of ornamental cast iron, the

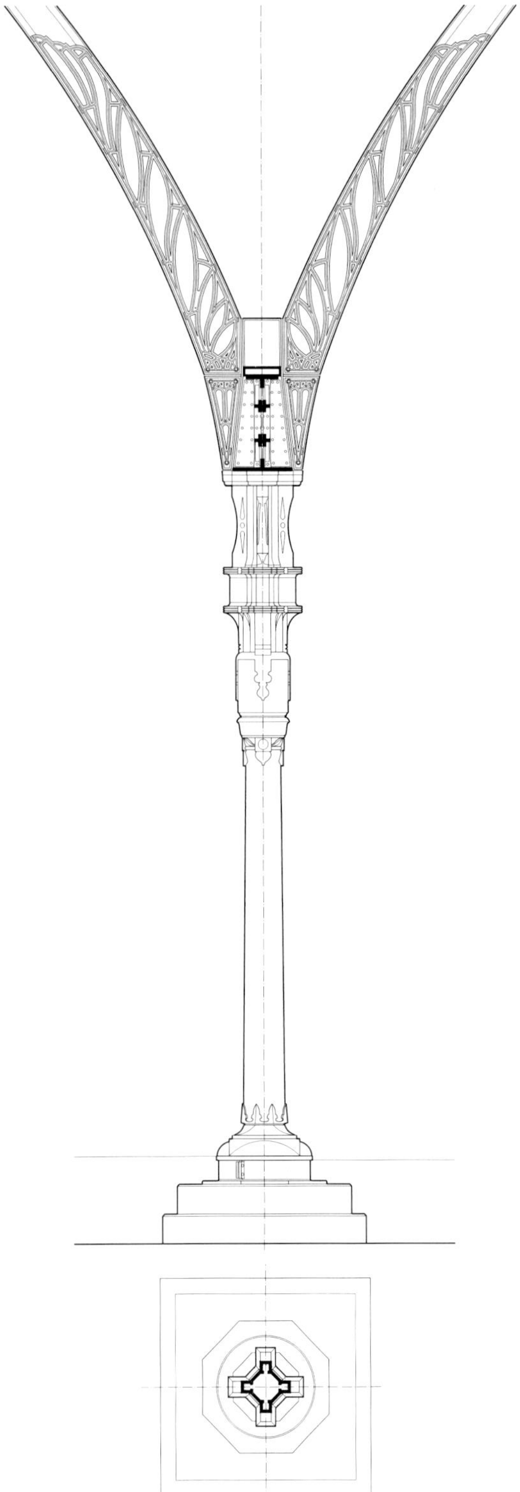

2 Longitudinal elevation of the columns at Paddington Station

pattern of Wyatt's columns was too strange and too personal ever to become a prototype.

Platform heights have been adjusted on several occasions since the station was completed in 1855. The largest columns were originally approximately 7.2 metres from their bases to the springing of the roof arches, and a sturdy 0.75 metres in diameter. Wyatt's cast-iron capitals took up over a third of their height. At the head, the capital branched out into ornate foliage panel-work, bolted to the face of the rib haunches, before merging into the wrought ironwork of the roof structure itself.

Each span of the station roof consists of sixty-three ribs at 3-metre centres, supported on twenty-one columns. Trusses, 9 metres in length, span between the column heads, each carrying a set of three ribs.

The columns also serve as rainwater downpipes. The iron gutters, which are themselves structural rafters, run along the valleys between the great roof arches, discharging every 9 metres into a concealed pipe that links to the column cylinder. The columns are fixed to the trusses at their head, and to their concrete bases, by means of simple mechanical bolts.

The rigidity of cast iron, and the mechanical assembly method, militated heavily against the approximation and fudge that was intrinsic to traditional building. In such a long, mechanically interconnected structure, dimensional creep would have had a devastating effect on the building process. To absorb minor inaccuracies in castings, which might otherwise have corrupted the setting out, details were designed to accommodate hardwood packing plates between metal members. The overall tolerance of the frame was controlled by planing or stacking the plates between the irons. It was for refinements of this kind that Brunel felt

3

4

5

3 One of the few surviving traces of Wyatt's columns; 4 Paddington Station, *c*1900; 5 Egyptian-style capitals in Owen Jones' *Grammar of Ornament*, 1856

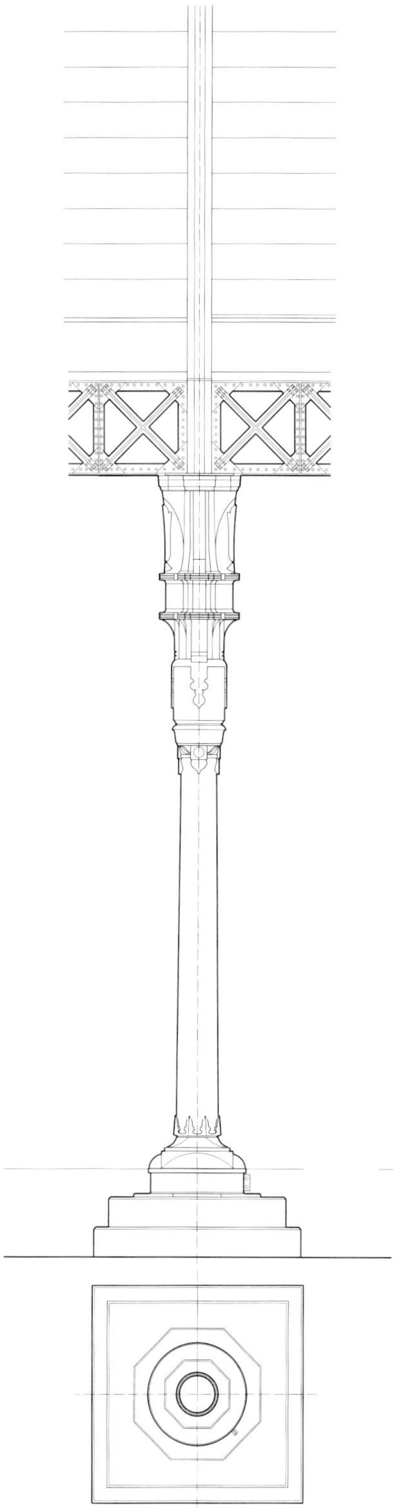

6 Transverse elevation of the columns at Paddington Station

compelled to contract Fox Henderson. The experience the firm had gained from working on the Crystal Palace had put it far ahead of any competitors, to the extent that the success of Paddington was seen to depend on its involvement. Fox Henderson was the only company invited to tender, and it retained the contract despite falling over two years behind schedule.

While Fox Henderson took control of the mechanics of the structure, Wyatt's role was limited, by Brunel, to its ornament. The design of the column capital shows the influence of Middle Eastern art, reflecting again the Victorian fascination with 'oriental' culture, an interest which led many of the country's better known artists to spend years of their lives wandering around the ruins of the south-eastern Mediterranean. Brunel's decoration consultant, Owen Jones (1809–74), was amongst these artistic pilgrims. His vision of modern design, (exemplified in his textbook *The Grammar of Ornament*, 1856), was characterised by a somewhat heavy-handed reworking of Arabic pattern. As a source of 'new aesthetics', ancient Egypt might seem today hardly to be an original inspiration, but in the 1850s, particularly so 'boldly' interpreted, it would have appeared daringly modern.

Brunel had commissioned Jones to make proposals for the decoration of Paddington with some emphasis on the colour scheme for the roof frame and columns. Wyatt's capitals were probably designed in liaison with Jones, which would account for their unselfconscious exuberance. If they had been made as intended, the shallow relief and perforated, blocky elements of Wyatt's abstract composition would have been differentiated by an equally confident colour scheme. Unfortunately, Jones's bold enhancements were firmly rejected by the Board of Great Western Railways, an organisation which, whilst having done more than any other to foster Brunel's genius, still often found his proposals too radical to stomach.

Brunel expected his collaborators to define their new architectural vocabulary in terms of ornament. The density and form of this ornament would distinguish the new aesthetic from the traditional. Whereas Philip Hardwick's lavish Great Western Hotel (1864), at the head of the station, was a typical Victorian picturesque concoction with elements of baroque, Renaissance and gothic, the ornament of the station ironwork was limited to star-shaped perforations in the roof ribs, simple linear paintwork, the proto-Art Nouveau wrought ironwork of the end windows, and Wyatt's cast-iron columns.

Many modifications have been made, and continue to be made, to the operational areas of the station. The large expansion and renewal works carried out in 1913–15 gave the GWR an opportunity to reassess Wyatt's columns. The cast iron that had been used in 1854 was, by that time, considered old-fashioned. Compared to wrought steel, it was brittle and unreliable. Rather than wait for failures to occur, the owners started the process of replacing them with wrought steel imitations. Although made 'in the manner' of the originals, they are really strictly second class passengers. Of all of Wyatt's original columns, at the time of writing, only one is known to remain, and that is beheaded. The originals may reside in storage somewhere along the Great Western rail network. It is to be hoped that future renewal work at Paddington, if it cannot actually restore some of the Wyatt columns, might at least revive the inventive exuberance of the station's first architects.

1 The Kish Bank Lighthouse, Dublin Bay

2 The Kish Bank Lighthouse under construction 'telescoped' down

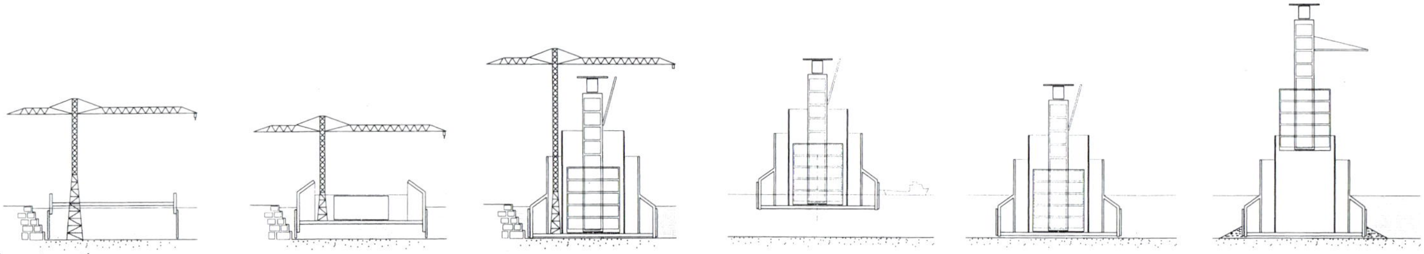

3 Construction sequence

CHRISTIANI + NIELSEN ENGINEERS

KISH BANK LIGHTHOUSE

Dublin Bay, Ireland (1963–65)

Nobody in their senses could claim it to be as beautiful a building [as the Eddystone archetype], but it is equally impressive in appearance and in technical achievement.
TG Wilson, *The Irish Lighthouse Service* (1968)

As previously discussed, columns have always been significant because they embody the higher instincts of mankind. Wilson is asserting that whether or not the modern Kish Bank lighthouse is conventionally beautiful, it is *meaningful*, and that, in the end, is the more telling virtue. Besides, it is an artefact of strange beauty.

The Kish Bank is a sand bar that, at low tide, lies near to the surface of the sea. For many years it was marked by a moored light ship. By the early 1960s, oil drilling, particularly in the Gulf of Mexico, had prompted the development of new techniques for raising tall structures at sea. These made use of the lifting capability of the sea water itself to raise massive steel or concrete components. The Commissioners for Irish Lights realised, at this time, that the new techniques might provide a way of raising a fixed lighthouse on the Kish Bank. A few small beacons had been built in Scandinavia, but the structure required for Kish Bank would be more than double the size of any earlier lighthouses made using the flotation-raising technique.

The Danish engineers Christiani and Nielsen were chosen to build the structure because they put forward the most economic and imaginative construction method. They proposed a 32-metre-diameter concrete caisson, to be sunk on to the seabed in harbour and drained, inside of which, in a second, tightly fitted concrete cylinder, the lighthouse would be built. When complete, the whole structure would be untethered from the harbour base, floated, and towed to the bank. Once in position, the caisson would be flooded to lift the column of the lighthouse into its elevated position, with the caisson below it acting as an anchor to the sandbank.

The plans went awry in December 1963, after completion of the outer caisson. A severe storm damaged it so badly that the builders scrapped it and used it as a platform upon which to build its replacement.

While work was proceeding in Dun Laoghaire harbour, tests on location showed that the sand of the seabed, itself below 18 to 22 metres of water, lay 90 metres deep over the rocks below Kish Bank. This confirmed that the weight of the column would not bear significantly on the bed rock, and that it would have to be carried by the sand alone. This was anticipated in the design of the wide, flat base which spread the load over the greatest practicable area. Rather than attempt to grade the sea-bed, gravel was laid to form a flat plinth over the sand.

The column also had to be substantial enough to resist the enormous pressure of hurricane waves. Climatic data suggested a once-in-a-century worst case of a wave 15 metres high and 150 metres long, propelled by a 125mph wind. A 1.60 scale model was tested under these conditions in a wave tank. Strain gauges were fixed to the model to provide the local design loads for each of the column's stages, and this data was, in turn, translated into specifications for the concrete and reinforcement design. As the components accumulated, the contractors began to be concerned about the build up of weight. The completed column would weigh 7,000 tonnes and, with the extra load of the old caisson, there would be too great a mass to float in the shallow water of the inner harbour. An intermediate location was found further out, and the telescoped vessel towed to the holding site in November 1964.

As work progressed, the builders had to maintain the balance of mass, buoyancy and sea depth, and for this reason the structure was again towed further out, to the limit of the harbour, in May 1965. From here there would be no need to wait for a particularly high tide to make the final journey. By 30 June the divers at Kish Bank had completed the preparation of the site. At dusk two powerful tugs took up the strain and, accompanied by a flotilla of small craft, started on the six mile journey to the sand bank. The following morning, the caisson was only a quarter of a mile from its intended site when a thick fog descended, obscuring the route markers. To add to the difficulties, flood waters unexpectedly increased the rising tide, and one of the tow cables snapped. A decision was made to postpone the final positioning, and the caisson was allowed to settle on to the sand bank where it came to rest at an ungainly angle, to the enjoyment of the sceptics who had haunted the project since its inception.

A few days later, however, conditions improved and the caisson was comfortably raised and resited in its proper position in water 21 metres deep. The compartments were then flooded and the whole unit settled slowly and carefully on to its prepared bed. The water in the outer compartment was replaced by sand, the inner compartment flooded, and the central column floated up 18 metres to its full height. The space under the tower was filled with sand and the annular space around it was packed with 30-millimetre aggregate and grouted to complete the concreting, adding an overall 18,000 tonnes to the weight of the structure. The fit-out was completed on the site, and the lighthouse handed over to its occupiers on 9 November 1965.

The top of the column, its 'capital', is equipped with a lantern containing a multicatadioptric apparatus, consisting of two faces each with four anodised reflectors revolving at 1 rpm, appearing from the sea to double-flash every 30 seconds. The light is 30 metres above sea level, and its 2,000,000 candle-power beam is visible for almost 20 miles. The 'abacus', above the capital, takes the form of a helicopter pad.

The lighthouse is equipped with a foghorn, radar and a radio beacon with a broadcast range of 50 miles. Internal generators provide power for all the signal apparatus and, for twenty-seven years, for the heating, cooking and other needs of the keepers, whose own quarters were designed to be compact, but luxurious compared to the dungeons suffered by their predecessors. Despite their relative comfort, however, even the keepers of Kish Bank eventually conceded to modern practice and abandoned the beacon to automatic control. The last keepers left in March 1992; the Kish Bank Lighthouse is still in continuous use.

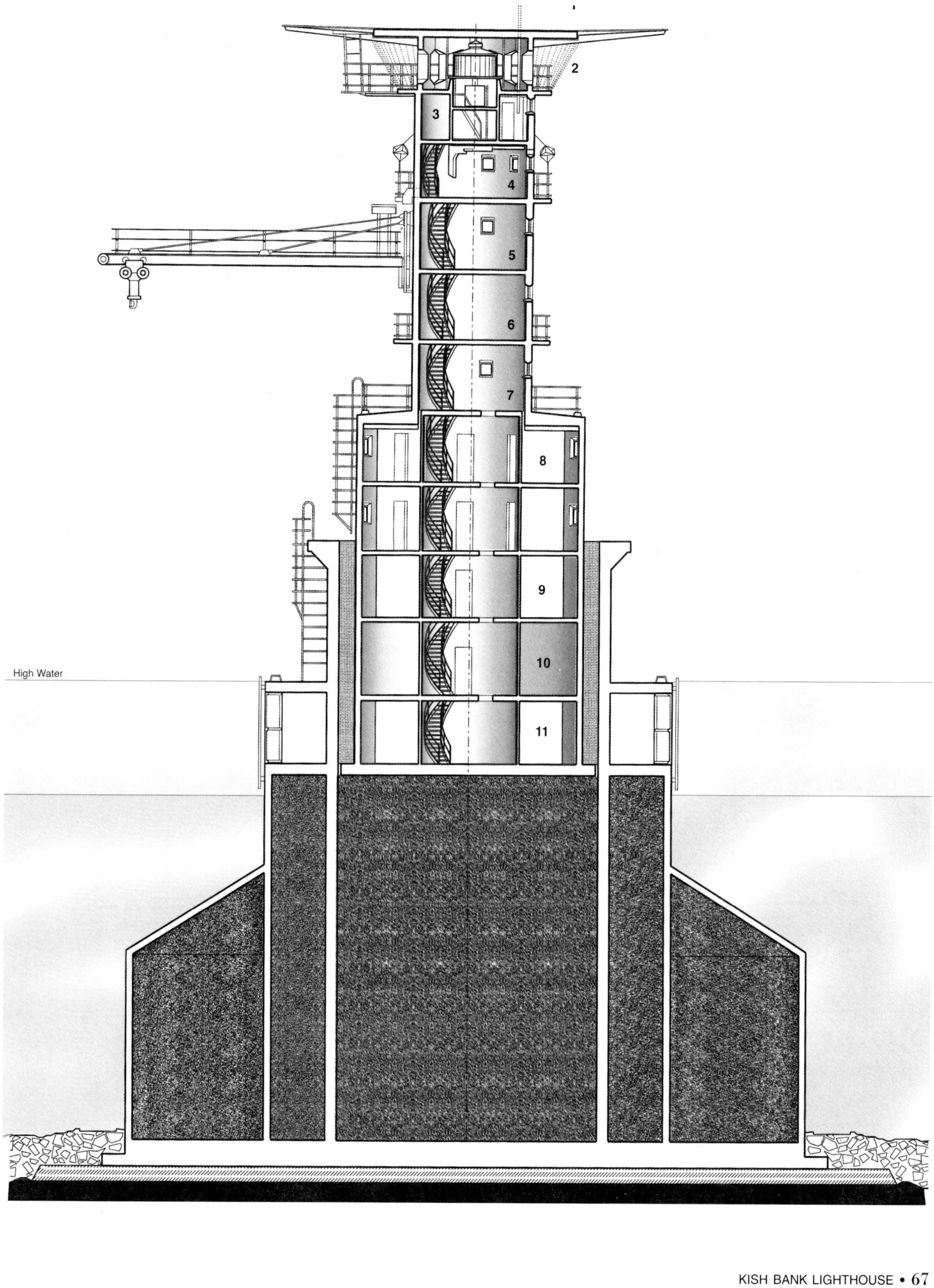

4 Cross section

1

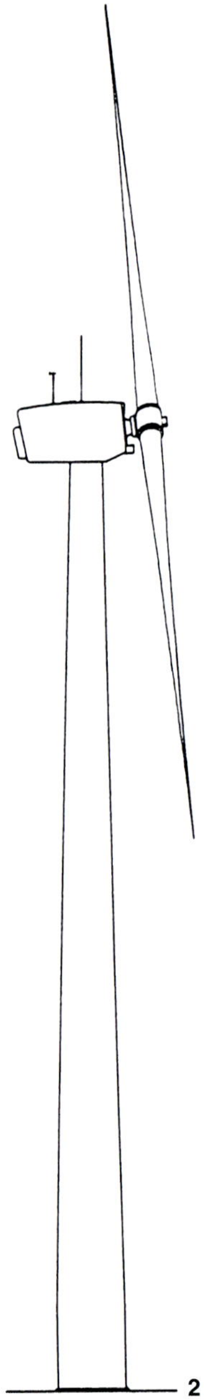

2

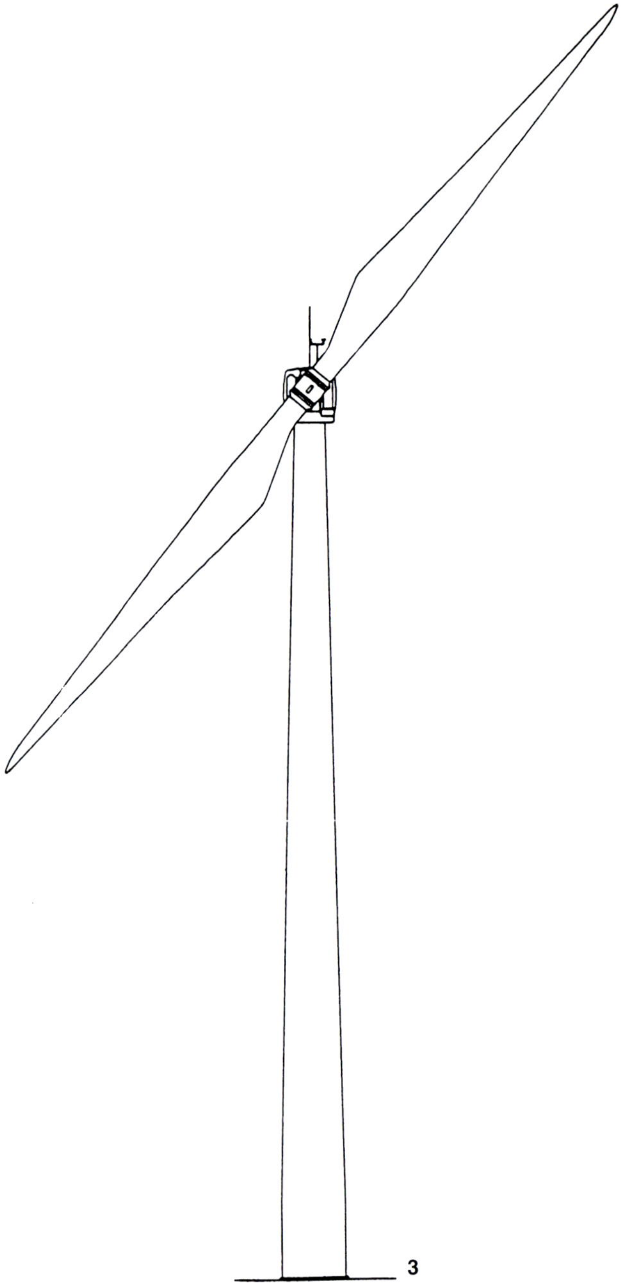

3

1 A three-blade wind turbine *in situ;* 2, 3 Side and front elevations of an MS3 wind turbine

WIND ENERGY GROUP

MS3 WIND TURBINES

1988–present

For any method of ecologically renewable power generation, the ratio of installation and operational cost against the value of usable power generated tends to be critically close to unity. In the optimum terrain and prevailing climate, wind turbines are among the few reliable means of extracting energy from the natural world without overtly polluting it in the process. Wind farms tend to function best in areas that are otherwise fairly wild, where most people would consider their presence to be intrusive, so the pressure is on the designers of the turbines to provide the strongest possible justification for their presence.

This is managed in two ways: firstly by emphasising, in the design, the properties of the structure and the turbine that allow it to generate power as efficiently as possible and, secondly, by making the structures beautiful. Whatever reservations we may consciously express, it is hard not to be stirred by the prospect of bare hills surmounted by clusters of propeller-headed white monoliths. In California and other parts of the world, where wind farms comprise not merely a few dozen turbine columns, but many hundreds, the Wind Energy Group's turbines have been found to be amongst the most efficient.

The largest turbine made by the company is the 3,000-kilowatt LS-1, built on the island of Orkney in 1987, and still capable of providing all of the island's power. It has a rotor 60 metres in diameter. The turbines in the MS series are designed to be arrayed rather than to stand as individual columns. The first of the series, the MS3-300, has an output of 300 kilowatts and was prototyped in 1988. There are now over a hundred in use around the world. The MS3-400, with a 39-metre rotor, is currently available, and the first large-scale installation of the 500-kilowatt upgrade is due early in 1998. The MS3-600, a 600-kilowatt turbine, will soon be available.

The MS3-500 has a teetered two-blade rotor. While three-blade rotors generally produce smoother power output in a wider range of wind conditions, they do put a great deal of stress upon the turbine. Two blades are considerably less stressful and, with some innovation in the design of the mechanism, the problems of irregularity in output are greatly reduced. On account of stresses on the nacelle – the 'capital' of the column – being so much lower, the housing can be much lighter. This makes them easier to lift, and a lighter gauge column is required to keep them aloft. Because ecology is the rationale behind these structures, lightness becomes a virtue in itself since, while not necessarily being cheaper, it implies less consumption of resources and a smaller environmental impact.

The rotor blades are produced from a high-performance composite of natural and synthetic material: birch wood veneers are bonded with glue layers of epoxy resin. The glue layers are reinforced with glass fibre and foam. The blade root attachment comprises thirty-six cast-steel pegs bonded into the birch composite matrix of the rotor. The profiling of the blades follows a NASA LS(1) aerofoil pattern, which is chosen to optimise the efficiency of the turbine. High and low wind speed versions of the rotor are made, with 39-metre and 45-metre diameters respectively, so that a wider range of site exposures can be allowed for. The 'teeter' mechanism allows the rotor to tilt up to 6.5 degrees into an axis where out-of-plane moments are reduced to zero at low speeds. This frees the rest of the rotor, nacelle and tower structure from the eccentric loading that would result if the rotor was unbalanced in the conventional way.

The power train components in the nacelle are mounted on a heavy steel bed plate. The nacelle housing is constructed from glass-reinforced polyester, and is carefully detailed to be weather resistant and relatively sound proof. The housing is 5.9 metres long and 2.3 metres high, with reasonable space to manoeuvre for maintenance inside. Access doors are installed in each surface.

The column is a simple conical cylinder, fabricated in seam-welded steel plate. Wind farm sites are typically fairly inaccessible, and this places constraints upon the size of components that can be handled. For this reason the column is transported to site in sections and bolted together *in situ* before erection. The diameter of the column is 2.2 metres at the base and 1 metre at the top. Flanges project from the cylinder at the top and bottom for bolting to the nacelle and the foundation. A fixed ladder is provided, with a fall-arrest device, which gives access to the nacelle. The whole structure has its own internal lighting, is entirely galvanised, and is supplied painted white.

The health and status of the turbine is monitored and controls adjusted as necessary by a computer which is housed in the base of the column. Access to the interior of the column is through a steel hatch at base level. The foundation is an octagonal concrete pad, 7 metres in diameter, and typically around 2 metres deep. A lightning-conductor earth cable is coiled around the base below ground level.

The nacelle and rotor weigh over 16 tonnes, and the column 24 tonnes. The structure is designed for simple erection and the system has been devised to require a single 130-tonne crane and a small crew of installers. Depending on accessibility and weather, once the foundation has been cast and cured, each column can be erected within a few days of delivery.

4 A pulley at the column head allows a mechanical ground-level winch to hoist the rotor and nacelle up into position; 5 An A-frame crane is used to lift the column into the vertical position about its hinged base; 6 Cross section through the nacelle of the MS3 wind turbine

4

5

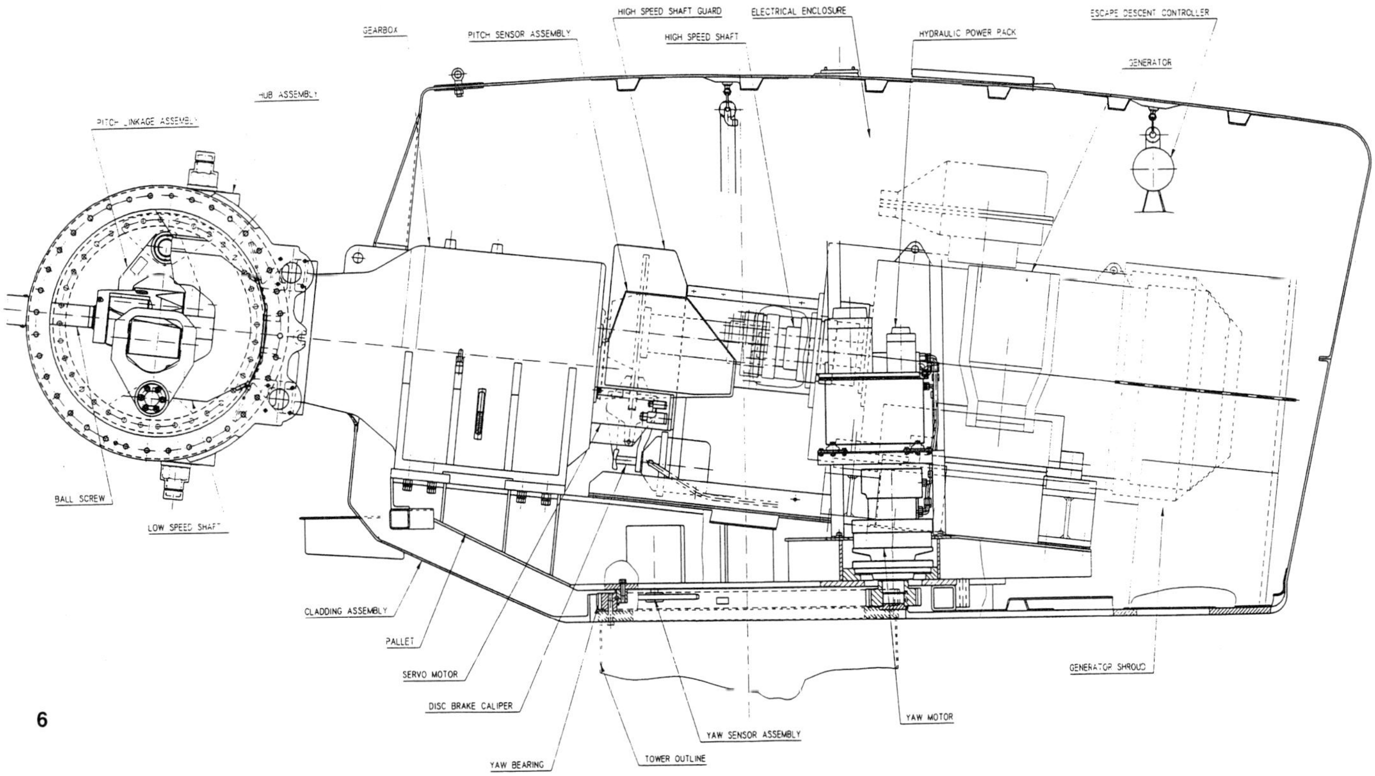

6

1

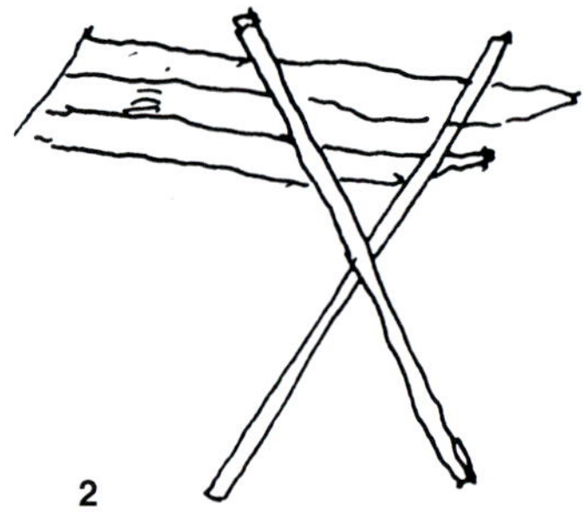

2

3

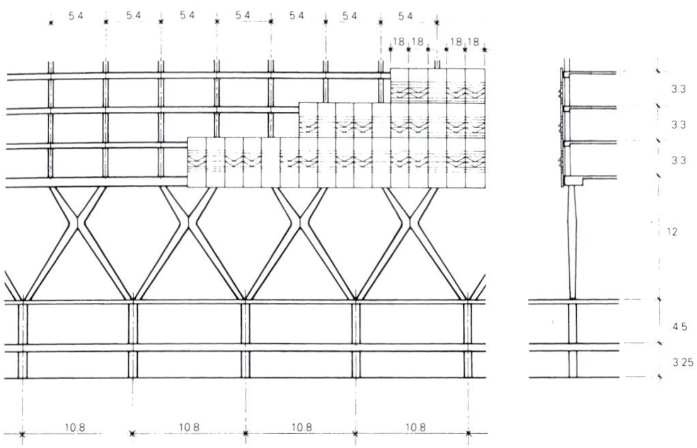

4

5

1 X-columns at the entrance to the Hôtel du Département; 2 William Alsop's concept sketch for the columns; 3 X-columns in the central atrium; 4 Section showing how the X-columns form a transfer structure which halves the grid between the below- and above-ground frames; 5 Colour tests for the epoxy coatings

ALSOP & STÖRMER

HÔTEL DU DÉPARTEMENT DES BOUCHES-DU-RHÔNE

Marseilles (1990-95)

The credibility of Post-Modernism was all but evaporated by the end of the 1980s. The few fresh specimens that have appeared since, such as Terry Farrell's MI5 building (1994) and James Stirling's No 1 Poultry (completed 1997), have tended to be delayed fruitions; they already appear out of their time.

Post-Modernism did direct attention to the design of columns, but none of the contemporary tendencies treat them as particularly significant. 'Modern' styles idealise the optimum clear span and often seek to eliminate columns in the interests of flexibility; minimalist design usually reduces its elements to planes; 'vernacular' thinks of columns as 'highbrow' and therefore intolerable. Fortunately, there are a few determined and influential designers who strive – occasionally above these conventions, and occasionally on their margins – to express themselves in building in a form that is recognisably distinct and their own. Amongst these individualists, William Alsop in particular has developed a varied and imaginative columnar repertoire.

The civic complex in Marseilles was a commission secured by Alsop in an international competition. As is typical of large competition-winning projects, the realisation of the building represents a considerable evolution of the initial proposal. Throughout the development, however, it remained the architect's objective to lift the massive prismatic volumes of the building off the ground plane so that they were clearly visible as resolved forms. In the earliest stages this was achieved by raising the building on simple, irregularly spaced verticals. As the engineers Ove Arup and Partners became involved, the three-storey high columns became identified as a means of reconciling the optimum basement structure (a loose grid to give wide spans within the basement utility zones) with the optimum office structure (a tighter grid to ensure that column sizes would be small enough for the walls to absorb, so as not to limit the flexibility of the floor space).

At one time, columns with flat dendriform branches were proposed for the transfer structure to divide this wide span into quarters. Ultimately, Alsop came up with the unique X columns. The spacing between the feet of the X is 10.8 metres but, because the two legs intersect at a point two-thirds up their height, the space between the tops of the legs is 5.4 metres, hence fulfilling the requirement of the transfer structure. The column design is a complex three-dimensional form, with delicate entasis and toroid concavity at the 'knee' joint. At the time, three-dimensional computer modelling and the engineer's finite element analysis programs were both still fairly novel as architectural tools, and this emerging technology was tested to the limit in the design of the X-columns before the details of their form and structural performance were adequately understood.

The logistics of execution can have a profound effect on architectural projects. The limitations of road transport in particular often critically influence design and fabrication decisions. (Where pre-fabrication and road haulage are seen as especially desirable, the width limits of vehicles often govern the form of major components.) In Alsop's design the X-columns dominate the public face of the building and a 'polished' concrete finish was essential – to take the high-gloss epoxy paint finish – as was a steel formwork fabricated to high tolerances. Precision concrete work of this kind is usually carried out in factory conditions, but at 12 by 10.8 metres the X-columns were too big to be transported. Any attempt to build up the columns from smaller pre-cast pieces would be bound to result in visible junction lines. A render coat could not be evenly enough applied. The only option was complete *in-situ* casting.

The polished steel shutterings were delivered in kit form and bolted together to make four-piece moulds; the left and right sides each had a back and a front. Although *in-situ* casting was the most economic technique to employ, the complex geometry of the columns significantly increased the risk of misalignment in the set-up of the mould. To minimise the danger of having to break and re-cast any of the massive columns, the shutterings were fixed into a travelling jig which was moved on precisely aligned railway tracks along the perimeter of the building.

Four rows of thirteen X-columns are used to support the two administrative blocks. At each column position, a site-fabricated reinforcing cage was craned into position in the open mould and anchored, the front of the mould was bolted into place, and finally concrete pumped into the mould through four funnel points and vibrated to ensure full compaction. The process, from assembly of the reinforcing cage to striking of the mould, took approximately two days for each column. The application of the paint (white in the atrium and blue on the exterior faces) took somewhat longer.

Alsop has already built more new column forms than most architects will dream up in a lifetime. The public swimming pool at Sheringham, Norfolk (1988), has columns that are three-dimensional ladder frames in timber; the Visitors' Centre in Cardiff Bay (1992) sits on top of splayed inverted tetrapod columns fabricated from sheet steel; the North Greenwich Underground Station, London (due for completion in 1998), has massive mosaic-coated struts of elliptical section, in a horizontal zigzag arrangement. Amongst his published but unbuilt projects, the multiplicity of 'leg' forms – some elegant, some ungainly – has become the architect's signature. In the more successful examples, as at Marseilles, the new column types need no justification other than that they befit the significance of the upright structural support in the formal composition of building and in the historical continuum of architecture.

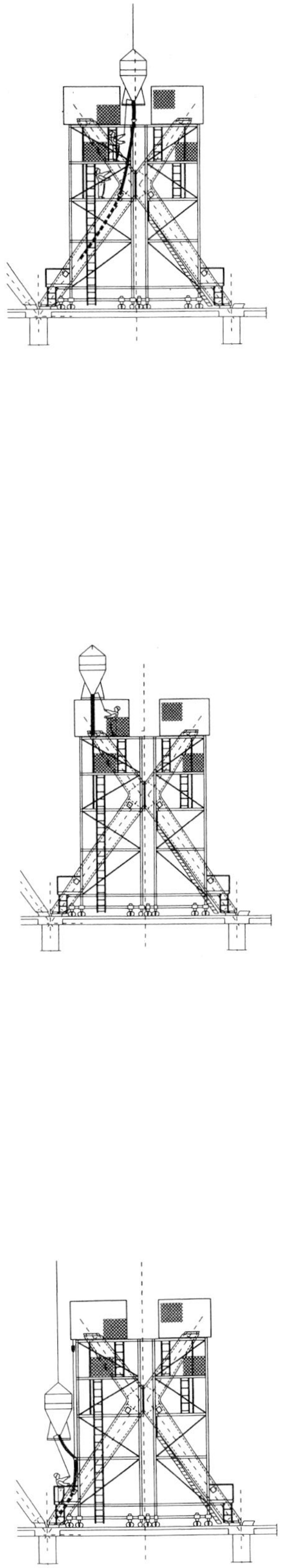

6

7

8

9

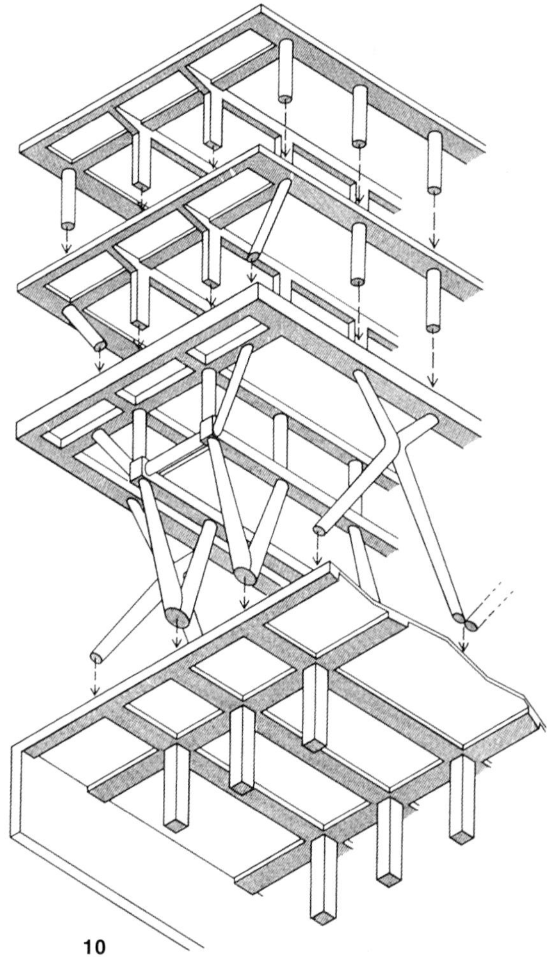

10

6 The concrete pouring procedure; 7 The transportable formwork erected for the first pair of X-columns; 8 Striking the shuttering of the first X-columns; 9 The floor slab over the completed colonnade being cast; 10 Worm's eye view showing complex structural hierarchy of X-columns, V-frames and diagonal braces at the end of the administration blocks

1

2

3

4

1 Troll GBS being towed out to its site in the North Sea; 2 Construction of the foundation skirts; 3 Construction of the lower column sections in the fjord; 4 Construction of the upper column sections

AKER MARITIME ENGINEERS (NORWAY)

TROLL GRAVITY BASE STRUCTURE

North Sea (1991–95)

In the late twentieth century there are very few building projects that cannot be rationalised in terms of economic advantage. This works in two ways: most often it prevents dreams from being realised; occasionally, however, it creates so powerful a vacuum that a quite extraordinary enterprise is required to fill it.

The largest offshore gas field in European waters is called Troll, and lies off the Norwegian coast in the North Sea. It contains enough gas to last for more than half a century, even if depleted at the rate of 84,000,000 cubic metres per day. The value of this output to Norway's consumers made it reasonable to build the Troll Gravity Base Structure, a drilling platform weighing 678,500 tonnes unballasted, and standing 472 metres high from its base to the tip of its flare boom. The columns that support the platform are, at 370 metres, the tallest vertical structural supports ever built. The platform, in an eerie echo of the Grand Menhir Brisé, is the tallest and probably the heaviest man-made artefact ever to be moved on the earth's surface. When ballasted for towing to its drill site, it displaced 1.25 million tonnes.

The dimensions of Troll GBS are determined by the depth of the North Sea over the Troll field. A floating platform would ordinarily have been used for waters 330 metres deep, but the fixed 'gravity base' became a feasible option, despite its staggering cost, because of the long anticipated life span of the reservoir.

The platform facility accounted for just over half of the £1.02 billion construction cost. This 'capital' is itself an extraordinary installation, built largely from titanium, aluminium and plastics composites to resist weathering. It is 170 metres long and weighs 22,500 tonnes. Another £500 million was expended on the drilling plant and concrete work below the platform.

The support structure was built in three phases. Phase one, the building of the foundation skirts, took place in a dry dock near Stavanger. The skirts form the base of the platform; when the structure is lowered to the seabed its weight submerges the skirt vessels into the soft clay to a depth of 30 metres. The nineteen skirt vessels are each 32 metres in diameter. Their covering domes are cast to the highest tolerances, on a reinforcement frame so dense that a tonne of steel is used for every cubic metre. This was necessary to withstand the pressure of 360 tonnes per square metre to which they would be subject during the submersion, and because, in order to avoid excess mass, the wall thickness of the vessels was limited to 500 millimetres. The spaces between bars and pre-stressing cables were so tight that conventional concrete was neither strong enough nor fluid enough to case the steel satisfactorily. This problem was solved first by developing a high-strength concrete formula and secondly by partly polishing the aggregate so that it flowed with a ball-bearing action.

The domes, which take the greatest pressure, also had to conform to the performance specification that no cracks should be larger than 0.02 millimetres. Despite their loading, this makes them, in effect, totally waterproof. These elements were cast by injecting the fluid concrete into the void between two skins of precast concrete. The skins were made in factory conditions to conform to the specification, and remained as the facing of the dome shells. By contrast, the walls of the skirts were cast, as were the other cylindrical elements, by slip forming. The overall area of the skirts is 1.66 hectares; the polished steel formwork itself was a massive structure, weighing 6,000 tons and lifted by over 1,500 hydraulic rams.

Once complete, the base was towed out into deep fjord water, where the remainder of the slip-formed structure could be cast by allowing the newly cast surfaces to sink gently below a floating formwork. The casting of the flotation caissons required marginally less precision, and to economise on weight as well as cost, pressure impregnated timber boards were used to construct the mould. Some of the caissons remain as evacuated chambers, and others are ballasted. For this reason, wall thickness varies between 500 and 1,500 millimetres.

The lower sections of the four shafts were cast on to four of the skirt chambers, symmetrically located among the caisson cells. This first 150-metre rise was cast at the first deep water site, using the special low-volume high-strength mix. The thickness of their walls varied from 0.6 to 1.5 metres as the anticipated pressure increased with sea depth. Two of the shafts are kept permanently dry. These buoyancy vessels also taper from 32 to 15 metres in diameter to reduce the weight acting on the base. Without careful variation of the concrete thickness, the platform would either be too heavy to avoid the sea bed on its 174-mile journey, or the shafts would be broken like eggshells by the implosion pressure. Although complex, it was still possible to achieve the variations of the geometry, thickness, radius and shape – from circular to elliptical – using the slip-forming technique that architects tend to think of as being limited to constant vertical extrusions.

The lower section extends up as far as the 'reigel'. This is both a knee brace and a stabilisation vessel. Conventional, smaller GBS platforms are floated into position by the buoyancy of caissons alone. Troll was too tall to be stable in the water without the second flotation vessel mid-way up the shafts. When towed, the top of the reigel tank is level with the sea surface. After building the reigel the structure was towed to its third and deepest construction site for the casting of the upper section of the leg and the 'topside' facility. The last 140 metres were cast using the same methods as the lower sections, but with wall thickness governed more by stiffness under wave loading, than the implosion factor.

The tops of the shafts are the first ever to be constructed with elastomeric bearings between the shaft and the topside structure. It is anticipated that the ability of the plastic inserts to absorb vibrations in the shaft structure will greatly increase the life-span of the drilling installation.

As huge as it is, it is tempting to see this structure as fundamentally quite simple. It is, after all, just four enormous columns. It is true that the physical phenomena of flotation and immersion are applied quite intuitively in construction, transportation and installation, but it was not easy to prove this intuition. As in the case of the Kish Bank Lighthouse, climate was

5

6

5 Photomontage of Troll towering over Oslo; 6 The submerged structure – note the foundation skirts driven by gravity deep into the sea bed

a critical consideration. The waves in the North Sea swell up to 30 metres, and the sea itself has ferocious currents. The conditions of towing presented many imponderables, as did the terrain of the sea-bed. These combined effects were hard enough to analyse; and then, at a very late stage in the design process, a previously unexplained physical phenomenon called 'ringing' came to the attention of the engineers, to complicate matters further. Ringing had been observed in large marine installations, and although not fully understood, it was generally not thought to be a danger. Troll was so large, however, that, as data on the phenomenon was collected, it came to be considered a potential threat to the integrity of the structure. The phenomenon is a random non-linear wave-induced dynamic effect; in other words, an occasional severe resonance caused by the impact of waves. To understand it, a 'random, irregular, non-linear time domain wave response analysis' is required. When the problem was identified, no such analysis existed. The structural design ultimately required what was almost certainly the most complex finite element analysis modelling ever undertaken in the design of a civil structure; the equation system incorporated nearly 850,000 unknowns. Their resolution required months of machination by one of the world's great number-crunchers, a CRAY Y-MP4D/464 Super Computer.

Troll GBS is never going to be seen; not, at least, before the seas run dry. It is a structure which can really only live in one's imagination. There is a means by which the platform crew can gain access to the interior of the 'dry' shafts, for maintenance. The prospect that greets them there – a flaring cylindrical shaft as deep as the Eiffel Tower is high – is unprecedented in construction, and surely a haunting experience. Aside from the imagination of such spaces, the only way one can attempt to appreciate the scale of this structure is from the endless list of superlatives that describe it: nineteen base domes, each wider than the dome of St Paul's; ten times the amount of steel, in reinforcement, as is used in the Eiffel Tower; over half a million tonnes of concrete cast to a tolerance of 2 millimetres and so on. It is, indeed, almost unimaginable. Its builders, Norwegian Contractors (now Aker Maritime) started building in 1991. The platform took less than four years to complete and it has been pumping gas to the mainland since April 1996. At today's consumer prices, the whole thing was paid for by the first few months of gas to be delivered. Perhaps it is because its existence is so easily accounted for in these terms that it seems not to qualify as one of mankind's most awesome creative achievements. In every other sense, it is a thing of wonder, and a fitting monument to the triumph of capitalist logic.

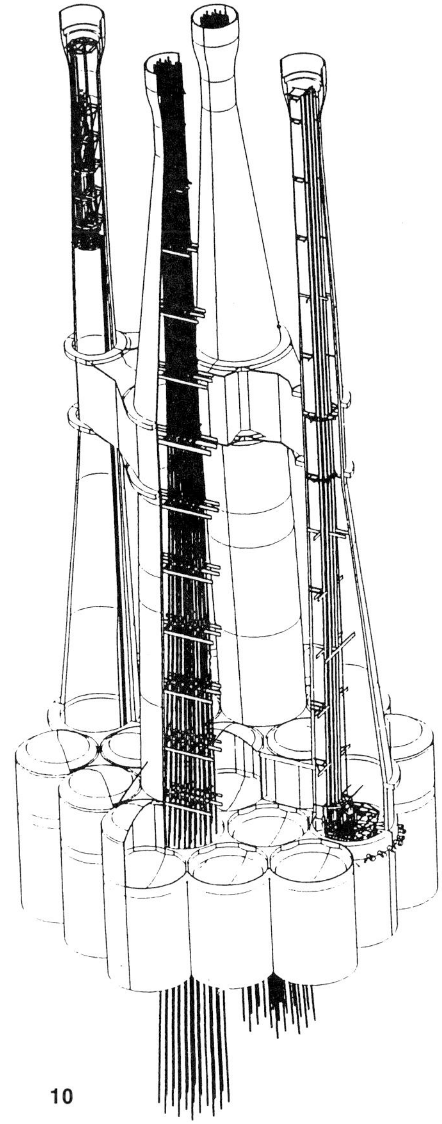

10

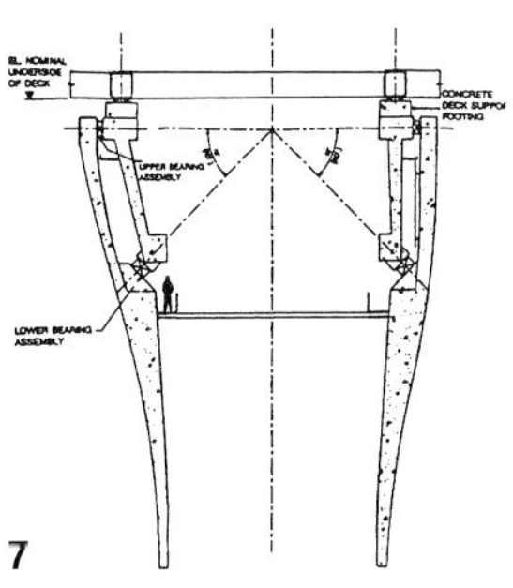

7

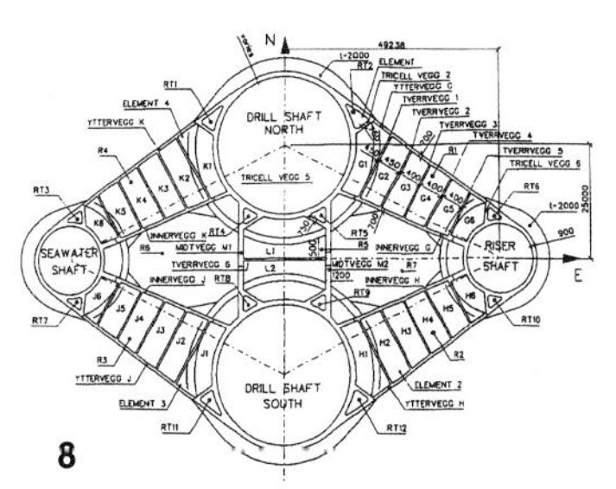

8

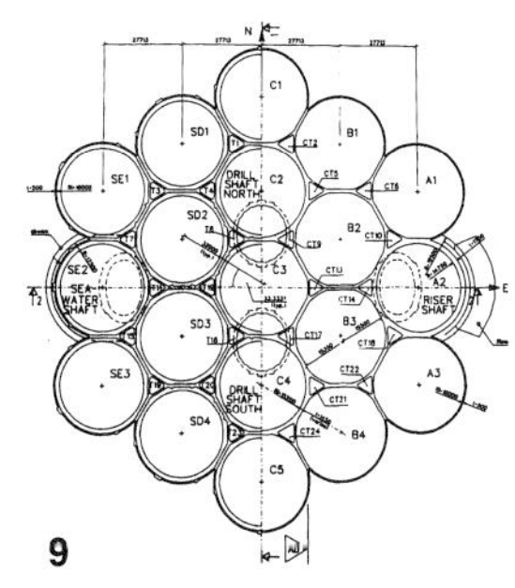

9

7 Cross section at column head showing elastomeric bearings; 8 Plan at the 'reigel' flotation tank; 9 Plan at the flotation skirt; 10 Cutaway isometric – note drilling equipment stored in 'empty' columns

1

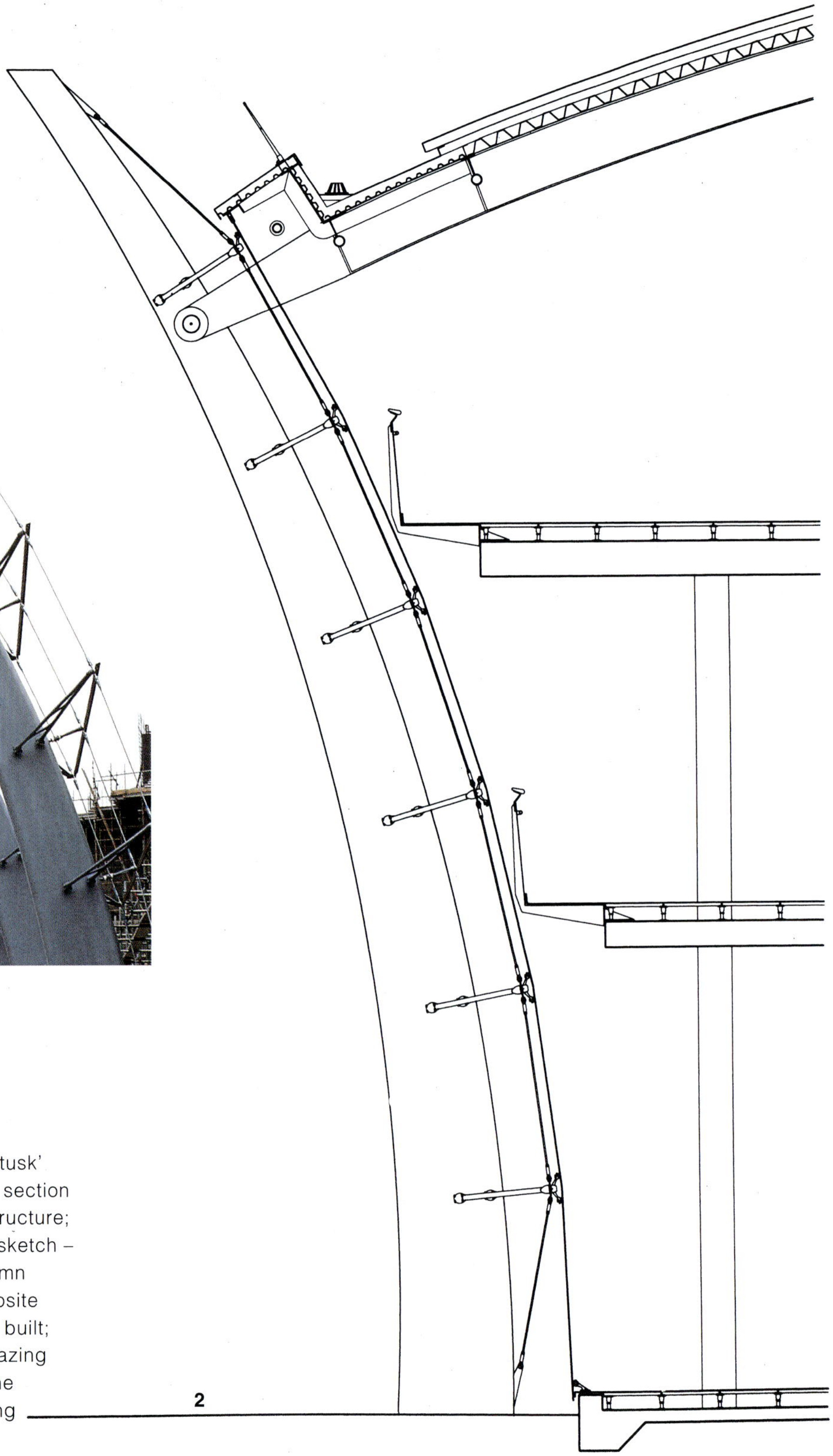

2

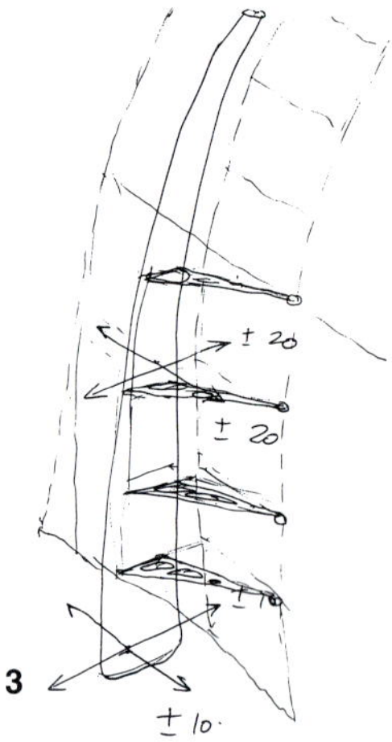

3

4

1 Erection of the 'tusk' columns; 2 Cross section of external wall structure; 3 The architect's sketch – note that the column bends in the opposite direction to those built; 4 The concave glazing and columns of the completed building

NICHOLAS GRIMSHAW AND PARTNERS

WESTERN MORNING NEWS

Plymouth (1992)

Until the late 1980s, the ex-partners Terry Farrell and Nicholas Grimshaw came to be identified inextricably with the two polar tendencies in British design: Post-Modernism and high-tech. By the end of that decade, Post-Modernism was in terminal decline as a credible ideology, and without its counterpoint, the high-tech orthodoxy quite suddenly appeared trite and formalistic.

Grimshaw's *Financial Times* plant in London was the last important building of the classical period of English high-tech. The most prominent features of the building are the vertical glazing supports that articulate its street facade. These are steel-fabricated fins, perfectly parallel and perpendicular. From each of these verticals a set of cast-steel branches reaches out to hold and locate the building's full-face glass wall. Behind the glass wall, as a secondary elevation, can be seen not serried floors of city workers, or specimens of primitive art, but a massive industrial newspaper press glinting and churning. Here, at the end of the classical period of high-tech, there was, for once, a comfortable synthesis of ornament and content; the building is a paean to its machine.

The fact that the *Financial Times* building attracted the unanimous acclaim of the popular press further fuelled anxieties that the genre needed either to evolve quickly, or risk withering into mere fad.

The *Western Morning News* building should not, therefore, have come as such a surprise. It was startling though, because it appeared as if, under the pressure to change, Grimshaw had revealed an unexpected sense of humour. This building, like any other high-tech design is, however, solidly founded on plausible rationale; the fact that it is ship-shaped is, according to its designer, a mere coincidence. The rationale concerns a bothersome, generally unnoticed feature of its direct ancestor, the *Financial Times* building. Whereas the massive printing press of the *FT* was spectacularly visible by night, during the day the glass curtain tended to reflect the sky, obscuring the architectonic bulk of the mighty machine. The transformative nature of glass, which had been celebrated (for example, in Foster Associates' seminal Willis Faber and Dumas building in Ipswich, 1975) was, in this case, decidedly unwelcome.

The *Western Morning News* building was Grimshaw's opportunity to amend the reflection problem. Sky reflection is most easily avoided by tilting the plane of the glass towards the viewer, so as to avoid gathering light from above. The higher up the viewer directs his vision, the greater must be the inclination of the reflecting plane towards him. Applied to an extended surface, this effect generates a concavity, with the viewer tending to be positioned near to the origin of the curve.

The shape of the *Western Morning News* building results from applying simple geometric optics to achieve the desired low reflection, and from setting out the perimeter line to follow the existing contours of the site. One can imagine the architect's delight when the familiar profile of a ship's prow emerged from this process.

As in the *Financial Times* building, internal floors are built off a conventional concrete frame. The internal columns are simple concrete cylinders. The extraordinary 'tusk' columns only support the glass walls and the roof.

The tusks are made from two hollow, circular steel sections, 160 millimetres in diameter, each rolled to a slightly different radius, with steel plates bridging between them on each side. The side plates are carefully inserted into the space between the tubes, and after a heavy build-up of seam welding, ground flush to be perfectly tangential to the surface of the circular sections. Whereas on the *FT* building the fin-like columns were straight, as if extruded, on the *Western Morning News* building they taper naturally from approximately 1.5 metres wide at the base to 0.5 metres at the tip. The inside of the curve of the columns is parallel to the curve of the glass wall. The column tip is cut at a slope and closed with a cover plate. They were delivered with their arms attached, one at a time, by road; at 15 metres high they were just manageable as single units. They were lifted into position by crane, using a removable hook plate that bolted on to the column tip, and fixed to individual concrete foundation pads by anchor bolts, just below ground level.

The glazing support arms are fixed to the column at approximate 2-metre vertical centres to align with the joints between panels of glass. The arms are a combination of mechanical ingenuity and almost Victorian ornamentalism. They are cast in spheroidal graphite steel by the ancient lost-wax process, from hand-carved patterns. The column shafts have tubular sockets, tapped with screw threads to receive bolts which are discreetly embedded into the terminal nodes of the arm. The fixing allows adjustments in tension between the column and the glazing plane, but does so without any outward emphasis of the joint mechanism. The pin joint near to the head of the column which carries the end of the roof beam is, by contrast, exaggerated. The beam ends are picked out in lighter paintwork, and the pin joint itself is capped on each side with polished stainless steel dishes.

Stainless steel is also used for the special glazing-boss castings, and for the network of tension rods that brace the ends of the support arms and tie back to the column at head and base. The tension rods keep the support arms aligned perpendicular to the curve of the column, and surprisingly, given their relative modesty, they also transfer most of the weight of the glass wall back to the column. The support arms, which appear the more purposeful elements, are actually pure wind bracing.

5

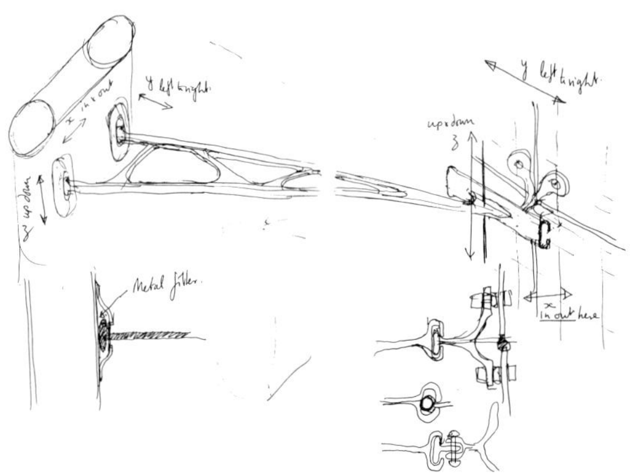

6

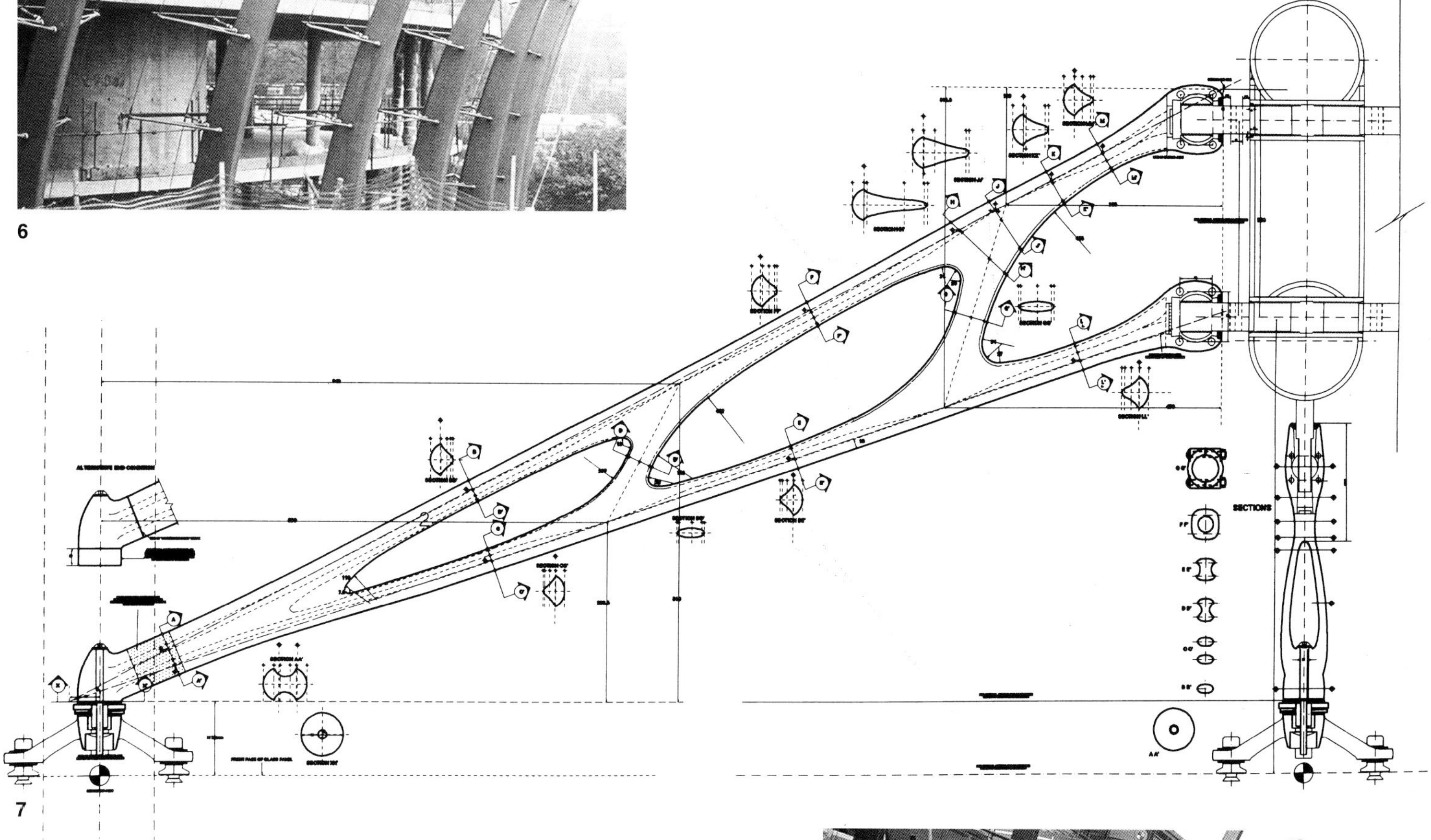

7

8

5 The architect's sketch of the glazing support arms; 6, 8 The wall structure under construction; 7 Detail drawing for the fabrication of the casting pattern for the support arms – the column is shown in plan

1

2

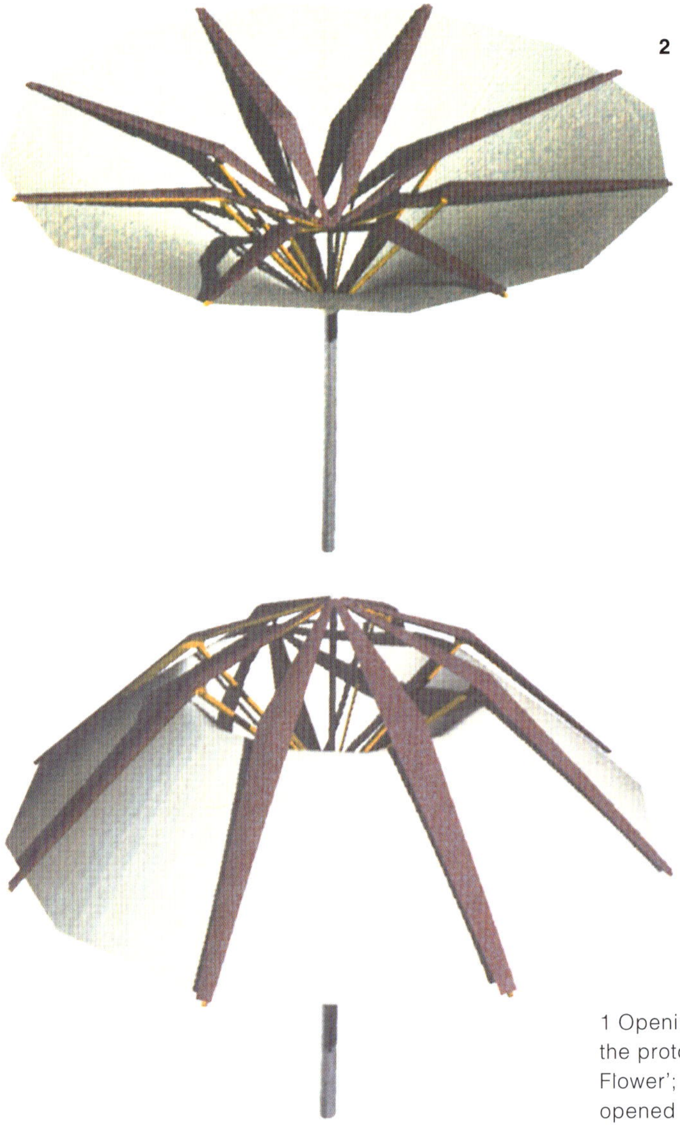

1 Opening sequence of the prototype 'Waddesdon Flower'; 2 The flower is opened by the telescopic movement of a central shaft into the supporting column; 3 Atelier One's initial proposal to fill a courtyard with many overlapping flower columns

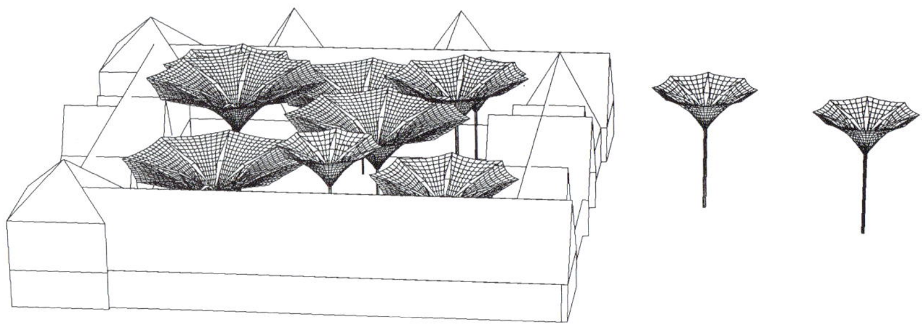

3

ATELIER ONE ENGINEERS

WADDESDON FLOWERS

Buckinghamshire (1993)

The imagery of mechanical movement is so intrinsic to modern design that architects, and their clients, can be forgiven for believing that it is a relatively straightforward matter to build the capability of true mechanical movement into a structural system. It is easy to overlook the fact that a system with integral movement must provide for a range of unpredictable structural conditions, and that the performance of motors is governed by the loads that they carry; as a structure moves the loads often change dramatically.

Two projects by Atelier One – the first for a freestanding column at Waddesdon Manor and the second for a structural support in the CableTel Exhibition Unit, represent what is probably the sensible limit of mechanical movement in structural design. It is certainly possible to build more complex kinetic structures, but because there is not usually an intention to mass produce, the time and expense of prototype testing and development is generally untenable within an architectural commission.

The umbrella-like column was designed for a restaurant courtyard at Waddesdon Manor. It is intended to be one of a cluster of similar columns which, when opened together, would shade the entire courtyard. The lower portion of the column is a tube, 115 millimetres in diameter, which acts as the housing for the telescopic action of the upper column shaft. The umbrella is raised by a motor fixed at the head of the lower tube. The motor has a worm-screw actuator movement that propels the upper tube in and out to raise and lower the umbrella. The drawing-in action pulls the crown of the column downwards so that the tips of the umbrella arms are raised by a see-saw movement, pivoting on the ends of the bracing struts. A series of sheet-steel collars and capping plates is employed to cover the mechanism, up-lighter and fabric canopy fixings. In principle, this is an elegant and economical mechanical system; the single linear action is the most efficient movement, and the worm-screw allows a high level of gearing and a smooth response.

However, the engineer was under pressure to use the smallest and cheapest motor capable of fitting unobtrusively on to the column stub and lifting and holding the 6.4-metre umbrella span. Almost inevitably this led to mechanical failure and redesign. The complete, functional 'flower' is in place at Waddesdon Manor, but the momentum of the project, which should have led to a spectacular field of flowers, was lost in the unavoidable process of mechanical refinement, and the finished flower is likely to remain a charming but solitary bloom.

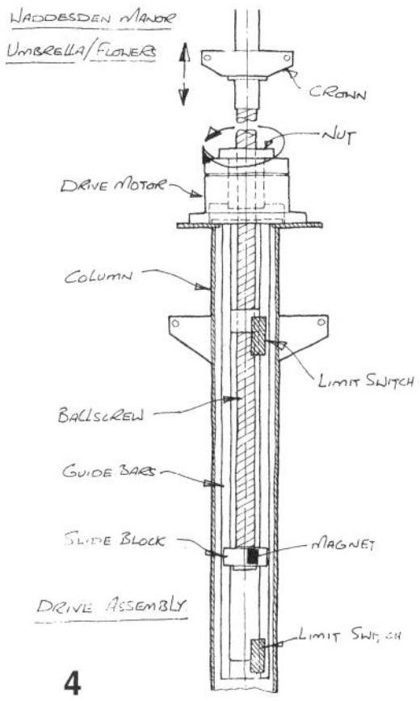

4

4 Section through lower portion of column showing drive assembly

CABLE-TEL EXHIBITION UNIT

(1996)

The Cable-Tel exhibition unit contains a travelling exhibition, and is designed to tolerate variable ground conditions, depending on its location. It is delivered on the back of a flat-bed trailer. On arrival, the unit's columns extend outwards and downwards mechanically to take the weight of the unit, allowing the trailer to drive away from underneath.

Once the trailer has moved out, the legs contract again, so that the unit is squatting just above ground level in its exhibition mode. The steel hinge bracket that anchors each set of fold-away legs is bolted through the internal frame of the exhibition unit. The sides of the exhibition unit taper inwards so that the legs can fold away within the width of the delivery vehicle. Each leg assembly comprises two pairs of pivoting rods which hold the top and the mid-span of the column foot. The feet on one side have a static plate, giving a friction bearing on the ground, while those on the opposite side contact the ground with wheels. These wheels allow the legs on each side to move apart when the exhibition unit lowers itself into ground position, and to be drawn together when the unit raises itself for transportation. The whole of the mechanical movement in each leg is generated by the telescopic action of a simple hydraulic ram. The four rams are powered simultaneously from a discreet compressor housed within the exhibition unit.

For delivering sheer power from very small plant hydraulic (or 'oildraulic') movements are still without equal in conventional engineering. Although they are powerful, the range of their movement is limited by the telescope length so it is usually more efficient to use the ram to generate a rotational movement, or lever arm, of the kind commonly applied to the jibs of cranes. This is the type of action used for the legs of the exhibition unit; a ram extension of 0.6 metres is employed, without gearing, to lift the unit nearly 2 metres off the ground.

Although it provides a powerful thrust, a ram should not be thought of as an effective structural element. It can carry a load, but under varying compression it deflects like a shock absorber. It is a stiff bar when contracted, but as it extends any significant lateral load will distort the cylinder and cause hydraulic fluid to leak at the mouth of the piston, which in turn causes a loss of hydraulic pressure and collapse; the further the ram extends, the weaker it becomes. For this reason when a ram is used to extend a vertical column, it is usually encased in a rectangular sectioned telescopic structural sleeve, so that lateral stiffness is achieved without loading the ram itself.

Like the Waddesdon Flowers, the development of the kinetic columns for the exhibition unit was protracted by testing and refinement. In this case, fortunately, a limited production run was planned from the outset, so the additional investment in mechanical design, although unwelcome to the exhibition designers, could be shown ultimately to make economic sense. No mechanical engineer can be expected to get every piece of a prototype system right first time; the whole point of a prototype is to reveal shortcomings. When an opportunity presents itself to incorporate mechanical movement into an architectural proposal, the

1

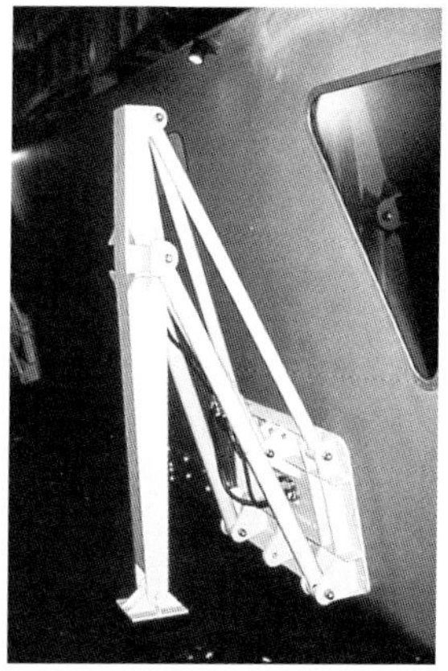

2

1 The exhibition unit raised on its columns; 2 The hydraulic column prefabricated and factory-fixed; 3, 4 The hydraulic column in lowered and raised positions; 5 The Cable-Tel exhibition unit is designed precisely to be carried on a standard flat-bed lorry

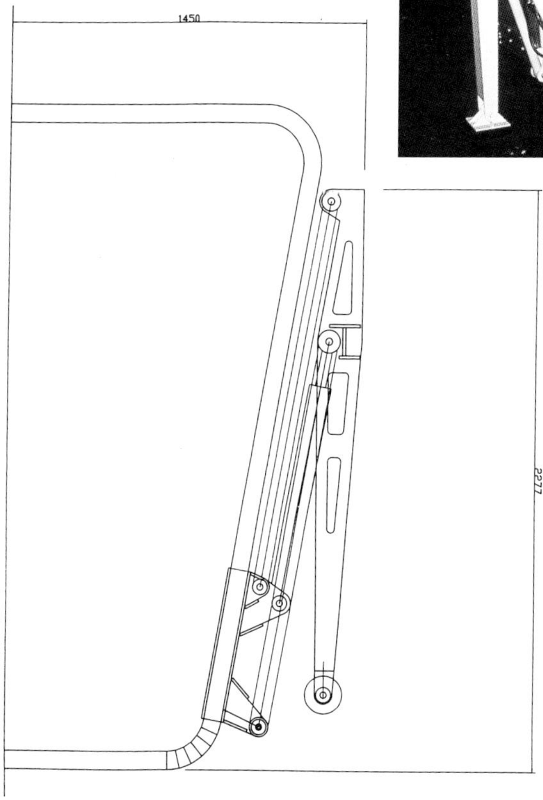

3

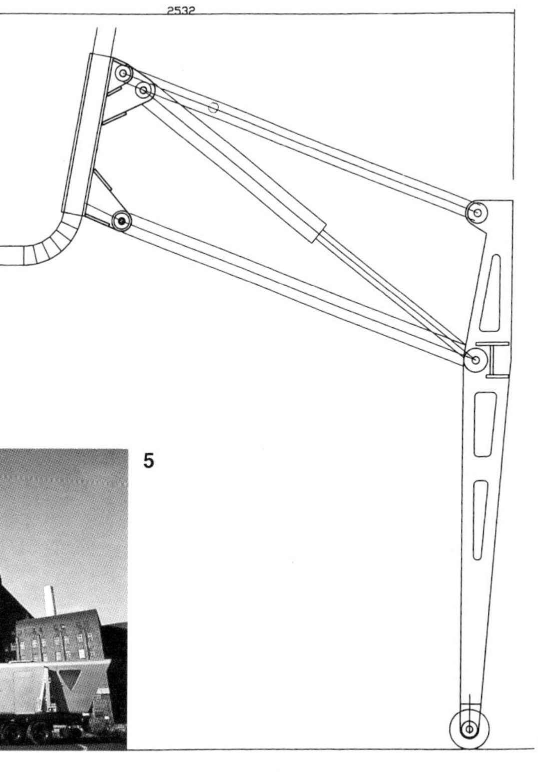

4

5

architect should remember that, as a general rule, every building is a prototype, and where mechanical movement is concerned, intuition is of limited use. Unless the project engineer has considerable experience in this area (and few have as much as Atelier One), mechanisms, however pleasing, should be handled with extreme caution.

1 Surge Column:
the Thames
Water Tower

BROOKS STACEY RANDALL FURSDON ARCHITECTS

SURGE COLUMN: THE THAMES WATER TOWER

London (1995)

This contemporary monolith is also an integral part of London's new 80-mile water ring main. It encloses one of three 15-metre high surge pipes which together accommodate pressure surges in the water circuit. Credit for the idea of turning this unprepossessing valve into a monument belongs to Royal College of Art students. The architects' realisation, which incorporates a number of advanced mechanisms is, however, as much a showcase for the concept of 'responsive' architecture as it is a celebration of the capital's water supply.

The column is designed with a concrete base, a transparent, polished shaft of radiused toughened glass around the core of the steel surge pipe, and a photo-voltaic cell array – a 'solar vane' – at the top as its capital. Whereas columns are traditionally thought of as having harmonious proportions – 1:6, 1:8 and so on – the Modernist ideal is simply to achieve the lowest possible ratio of girth to height. The more classically proportioned clear shell and the slender solid steel core of the 16-metre high surge column give it two effective silhouettes, depending on the lighting conditions, and the design of the column is developed to make a feature of the proportional dialectic.

Presumably, because the static form was considered insufficiently engaging for a truly contemporary monument, the architects have built into it a system of sensors, triggers and water jets that reacts to changes in atmospheric air pressure: the column is also a barometer. The power to operate the barometer mechanism is converted from sunlight through the photovoltaic cell array at the top.

The internal space between the steel pipe and glass shell is divided into five vertical portions. The array of spray nozzles that jet blue dyed water on to the back face of the glass shell is mounted on the steel core. If the ambient air pressure is very low – which in London would be likely to indicate stormy weather – the blue water is squirted on to the lowest portion of the inside of the glass cylinder. If the skies are clear and the atmospheric pressure appropriately high, then blue water is sprayed into the cylinder all the way up the five portions. Most of the time, unsurprisingly, the blue level hovers around half way up.

The blue-dyed water is one of the principal visible elements of the column. Around 500 litres are required to fill the system, and the fluid itself is carefully engineered for the desired viscosity, anti-freeze and anti-algae characteristics; even UV stability, as blue pigments are generally prone to sun-bleaching. The matt-finish stainless-steel enclosure in the column's base contains the water circulation plant. Above this zone the surge pipe itself is clad in mirrored stainless steel.

Branching off the sturdy steel walls of the surge pipe is the network of sand-cast steel armatures that supports the glass outer shell.

The horizontal glazing divisions align again with the five barometer stages, and black silicone mastic is used to seal the horizontal joints as a concession to calibration. The rungs of a fixed maintenance ladder are also intended to be read as a gauge for the barometer. However, without blatant signage, it is unlikely that many of those whose eye it catches, as they crawl around the tormenting bottleneck of the Shepherd's Bush roundabout, realise that they are looking at a barometer. Some assume that it is purely a utility installation, some that it is a piece of architecture, and others that it must be sculpture, and the column benefits hugely from this tantalising ambiguity.

The architects describe their column as a working model of a responsive building, designed to create a visually poetic effect. Many contemporary architects and engineers have expressed excitement at the potential of automatic sensor controlled environmental systems – 'intelligent buildings' – but few have been able to find ways in which systems of this kind can enhance the actual architecture. For various reasons the allure of intelligent building will become irresistible and its formal genre will, in due course, take shape. In the meantime, the architects of this flamboyant modern monolith have been able to show that dry technology can indeed be moistly poetic.

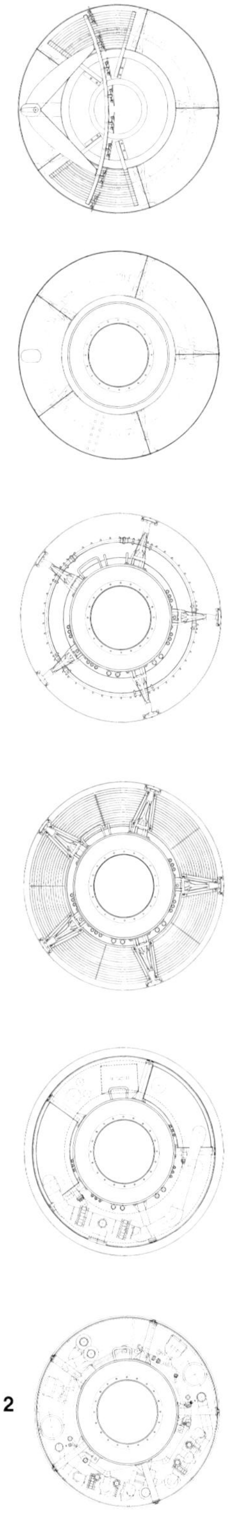

2

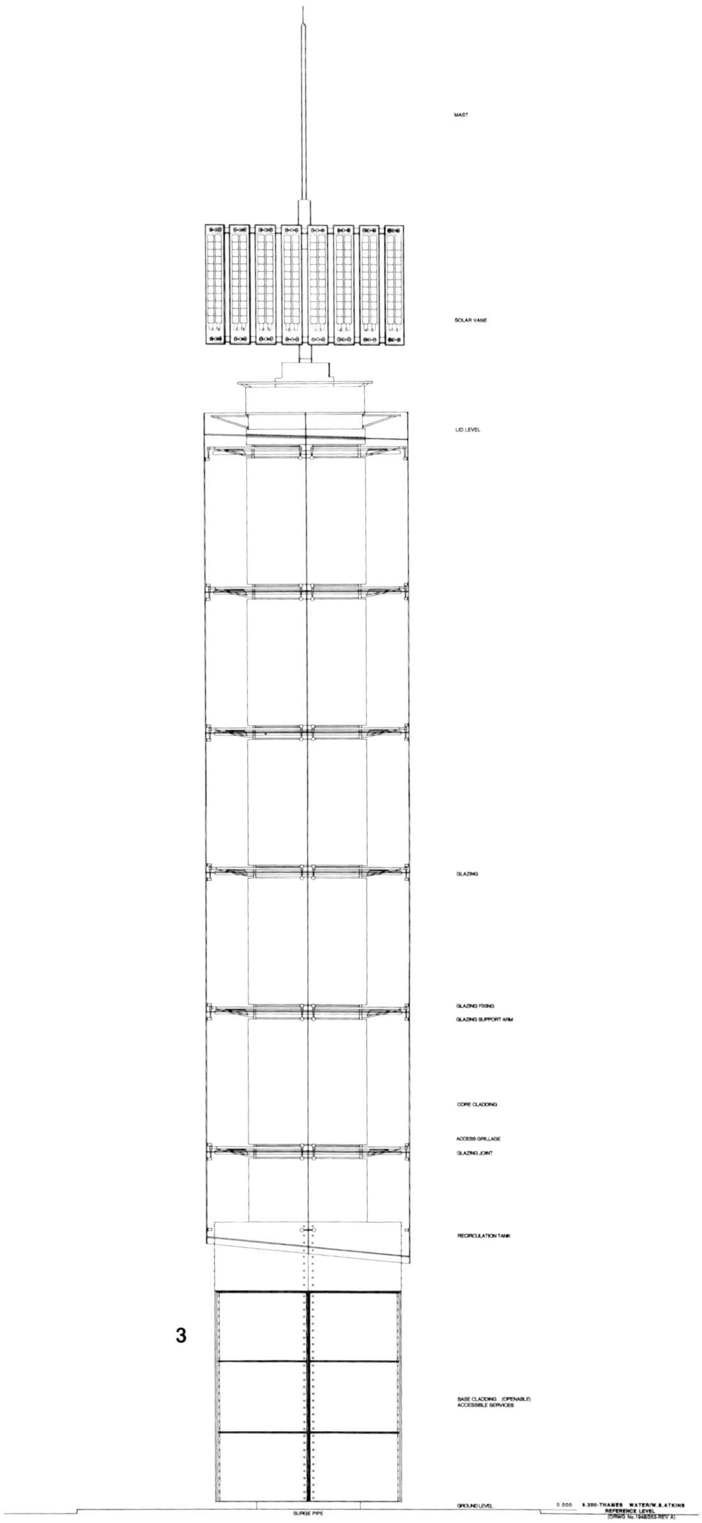

3

2 Plans showing, from above: solar vane level; top of glazing; water distribution system; typical access level; recirculation tank level; core cladding level; 3 South elevation

4

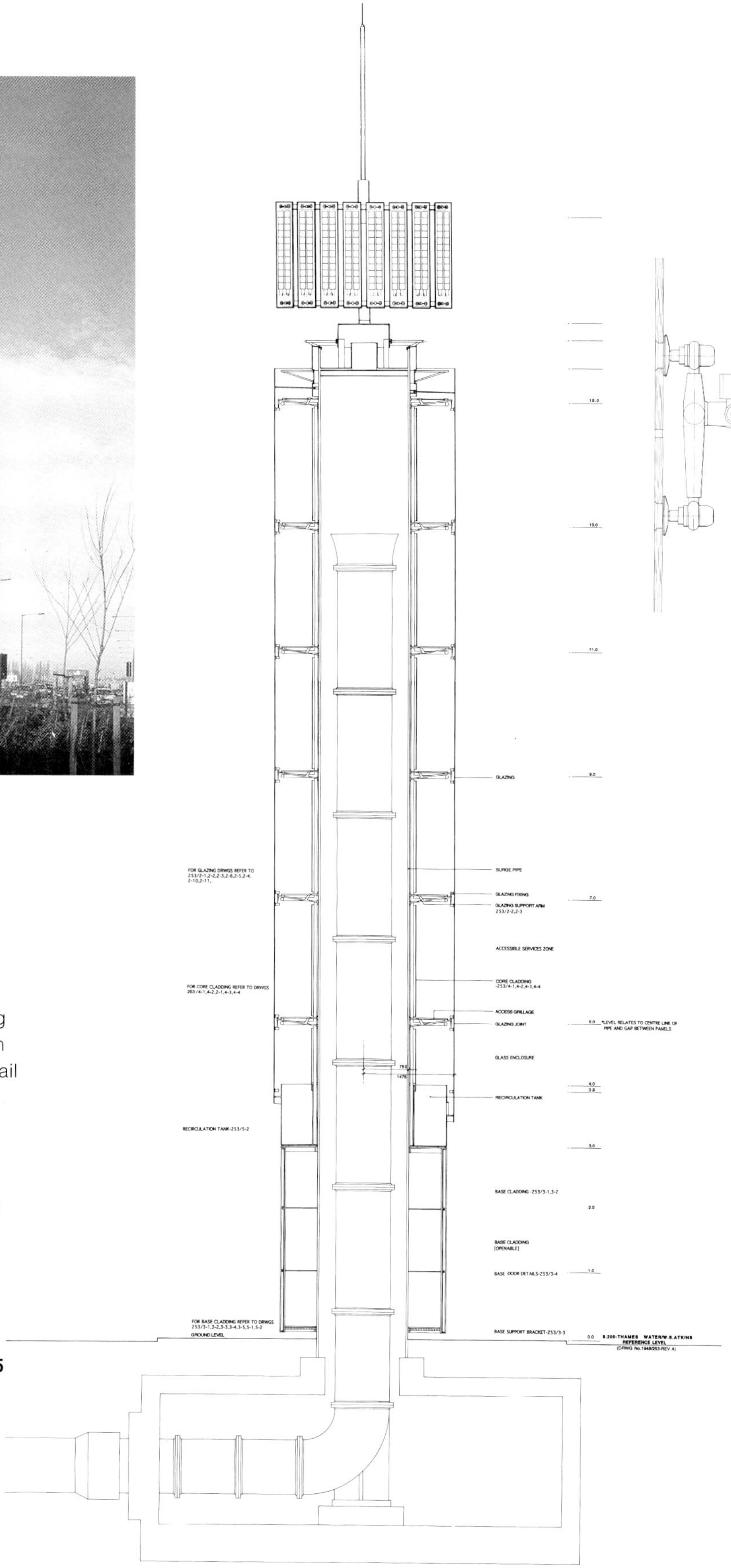

5

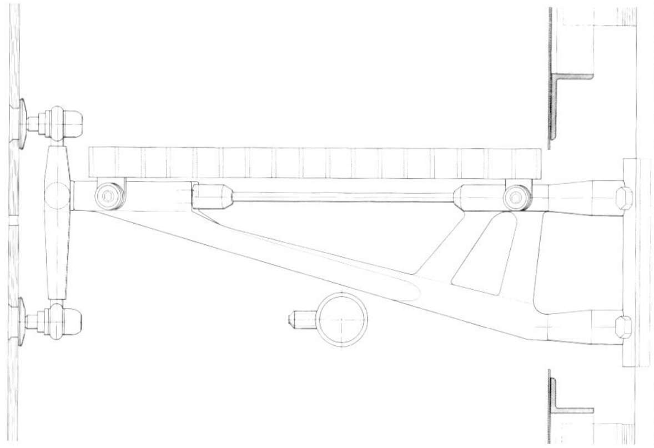

6

4 The blue water of the Surge Column indicating a temperate afternoon in London; 5 Section; 6 Detail of casting supporting grillage and glazing; 7 The Surge Column as landmark amidst the traffic of the Shepherd's Bush roundabout

7

1 Interior of the Judge Institute

JOHN OUTRAM

JUDGE INSTITUTE

Cambridge (1995)

John Outram is another of those rare creative architects who has acquired an overview of architectural history, and who has interpreted it in a manner that is purely personal. This project for Cambridge University, a college faculty of management studies, contains echoes of Post-Modernist formalism; however, while its classical references are overt, the supporting rationale is quite distinct, modernistic and entirely at odds with Venturian theory.

Architecture can and should mean something; classical form is *per se* a vehicle for meaning; and architecture is fundamentally bound to classical form in the same way that painting, however abstract, can never be separated from representation. The assertiveness and functionality of the great cylindrical verticals in Outram's building suggest, firstly, that the architect believes columns should be pre-eminent among architectural elements and, furthermore, that the rest of the antique hierarchy of classical ornament continues to have relevance in contemporary design *provided* that it upholds contemporary moral values. (Post-Modernism, on the other hand, is distinguished by its *amoral* superficiality.)

The least controversial moral assumption in contemporary architecture is that buildings should be energy efficient, renewable and generally 'green'. The column-lined atrium of the Judge Institute Gallery is 24 metres high from ground level to eaves. The atrium space is the fulcrum of a building management system that utilises the structural mass (as a 'heat sink') and high-level ventilation openings (for the atrium 'chimney effect') to augment the mechanical heating and ventilation of the building: the whole of the gallery is a 'green' machine. The columns are structural, but not solid masonry; an armature of reinforced concrete members girdles each of the columns, picked out in coloured paintwork like belts and braces. The brick or blockwork shafts are straightforward infill.

Outram is, with some justification, disparaging about the formulaic core-centred design of modern work places. Cores are lift shafts that have evolved to become the service trunk-lines of modern buildings. Certain engineers would claim that their design is almost a science in itself. The Judge Institute has no cores; all the vertical service runs are dispersed from central basement plant rooms upwards through the multiple-rising column shafts. Some of the columns are dedicated to utilities trunking – single- and three-phase electrical conduits, local area network cabling, telephone lines, drainage and plumbing – others are vertical air handling ducts, with funnel like capitals which are, for once, actually funnels, drawing air from the underside of the gallery roof back into the ventilation circuit. Environmental sensors, which measure rainfall, wind direction, temperature, and levels of carbon dioxide and light are also embedded in the columns. These monitoring

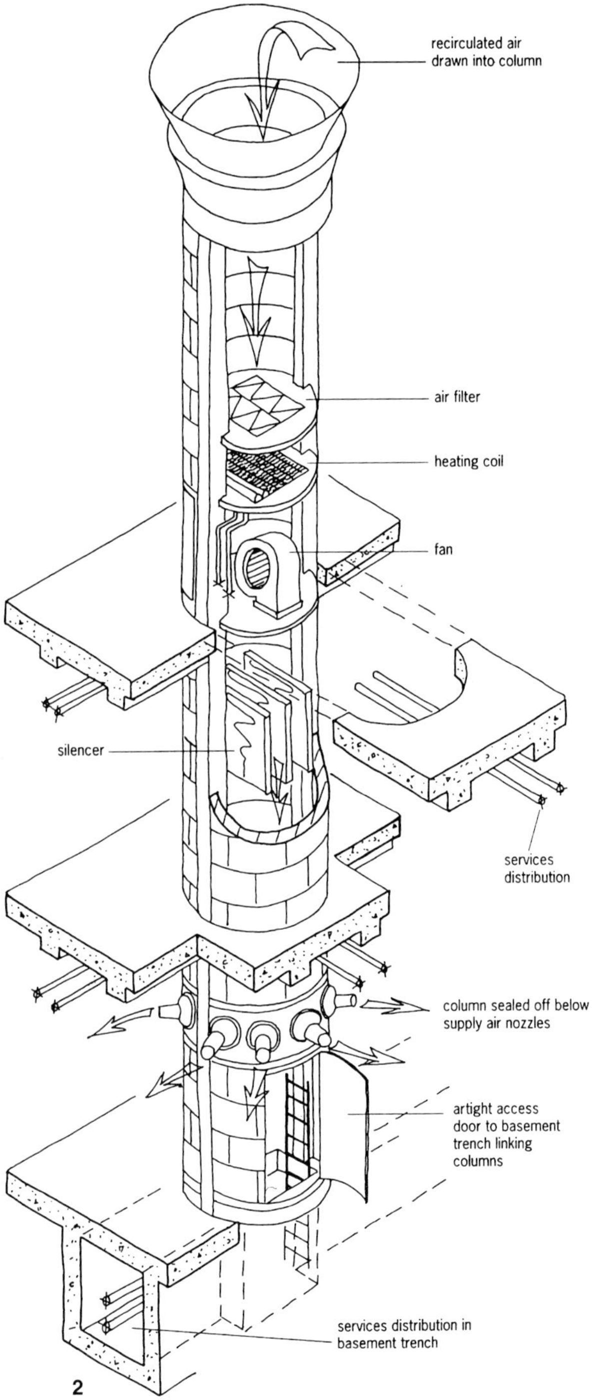

2

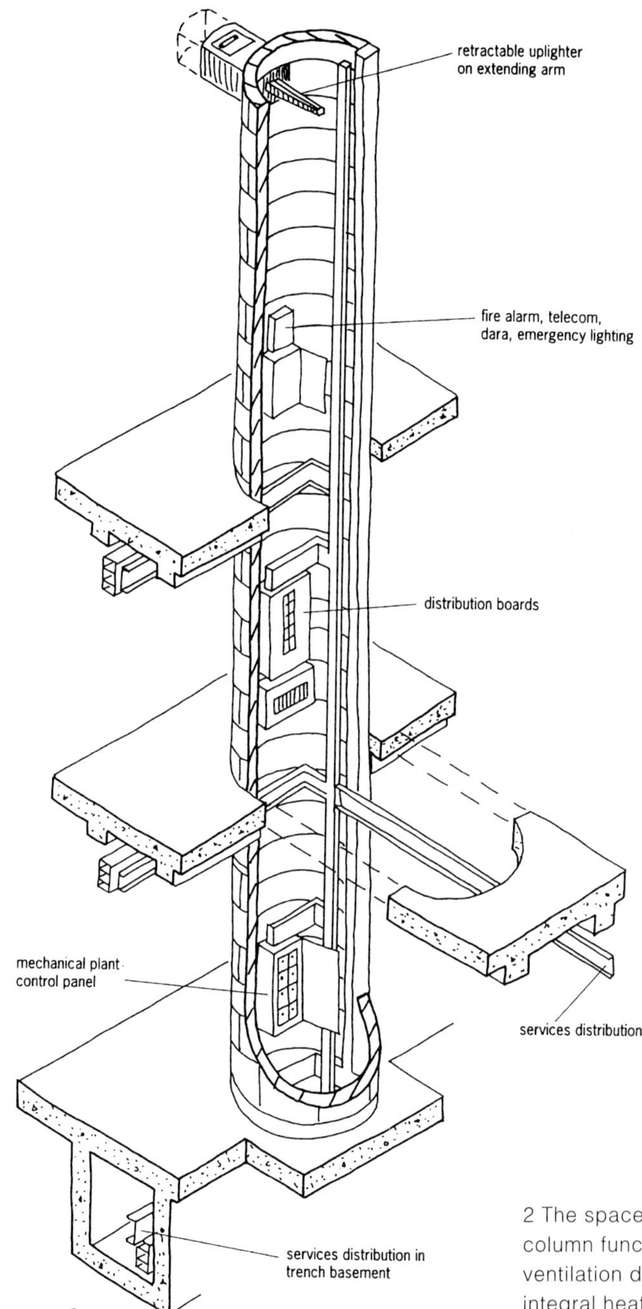

3

2 The space inside the column functions as a ventilation duct with integral heating and filtration; 3 The space also functions as a service core housing all the IT and environmental control apparatus

devices feed data into the building management system, which makes automatic adjustments to the vents, fans, heaters and so on in order to maintain a constant environment inside the building. Apart from the tubular silver air nozzles, other outlets, sockets and grilles are camouflaged to match the blockwork of the columns. This is even true of the recessed metal-halide light fittings which slide out automatically from the columns at 17 metres above floor level, to project light up to the underside of the roof when ambient light levels fall. Service branches emerge from the columns level with the underside of the ceiling of each intermediate floor. One laudable objective of the system is to do everything possible to ensure that suspended ceilings are never needed. The concentration of services inside the columns also means that maintenance can be carried out inside the cylinders, without obstructing the use of the building. Curved doors are provided at every floor level for maintenance access to the column interiors.

While Outram has the courage to use the colour available to him, he still does so by reference to the precedent of 'polychromy', as if the forceful use of colour is not after all a natural prerogative of the visual artist. 'Polychromy' is a euphemism. Like all euphemisms, it is used to sweeten an otherwise unpalatable fact; in this case, the fact that the virginal white columns of ancient Greek temples, an image so precious to Beaux-Arts classicism, were originally painted in colours that we would consider garish and vulgar. For centuries classical revivalists have ignored this. The columns of the Judge Institute are, however, embellished with a variety of coloured and patterned concrete blocks. Outram has proclaimed that:

> Concrete is my material of choice because it is free to assume any colour, form or pattern . . . it offers no easy paths to form that can be attributed to external imperatives such as gravity or nature.

Outram, like Alsop, is of that rare breed of architects who believe the public is best served by forthright expression, and that those of his contemporaries who cling to 'constraints' do so in order to avoid having to think what they really want to say.

Having cleared the ground in this way, Outram reaches back to the classical repertoire in order to find some structure of meaning, and then invokes the politics of ecology in order to find a structure of morality. As creditable as they may appear, the architect's true motives for incorporating 'green' principles in to building are as arbitrary and technology-driven as any other programmatic ideal. He has even labelled his columns as the Robot Order, to emphasise their functional validity. It is difficult to avoid sensing a paradox in that an architect so assertive of his personal expression, and so dismissive of functionalist rationale, should resort to a 'green' programme to substantiate a visionary, gigantic classicism that should need no justification, other than that it is magnificent.

4

5

4 'Robot order' columns lining the 24-metre high central atrium; 6 Concrete structural girdle and ventilation nozzles of the columns

BIBLIOGRAPHY

Primary Sources

Vitruvius, Marcus, *De architettura*, c40BC: Various Latin copies of the manuscript survive which differ quite significantly; the two most popular English translations are based on different versions: Granger, F (ed and trans), *On Architecture*, 2 vols (London and Cambridge, Mass, 1931, 1934); and Morgan, MH (trans), *The Ten Books on Architecture*, Harvard University Press (Cambridge, Mass) 1914; reprinted by Dover Publications in 1960.

Alberti, Leon Battista, *De Re Aedificatoria* (Florence), 1485; English translation by J Rykwert, N Leach and R Tavernor, *On the Art of Building in Ten Books* (Cambridge and London), 1988.

Serlio, Sebastiano, *Regole generale di architettura sopra le cinque maniere de gli edifici cioè Thoscano, Dorico, Ionico, Corinthio, et Composito* (Book IV), (Venice), 1537; English translation of books 1–V by AE Santaniello (New York), 1980.

Palladio, Andrea, *I Quattro Libri dell'Architettura* (Venice), 1570; English translation I Ware (London), 1738.

Fréart de Chambray, Roland, *Parallèle de l'Architecture Antique et de la Moderne* (Paris), 1650; English edition by John Evelyn, 1664.

Vignola, Jacopo, *Regola delli Cinque Ordini d'Architettura* (Rome), 1562; first English edition, 1669.

Jones, Owen, *The Grammar of Ornament*, 1856; reprinted by Van Nostrand Rheinhold (New York), 1982.

Ruskin, John, *The Seven Lamps of Architecture* (London), 1849; reprinted by Century (London), 1988.

Le Corbusier (Rodker, J, trans), *Towards a New Architecture*, Architectural Press (London), 1927.

Further Reading

AA Files, no 21, Architectural Association (London), Spring 1991.

Ackerman, James, *Palladio*, Penguin (Harmondsworth), 1966.

Appleyard, Brian, *Richard Rogers: A Biography*, Faber and Faber (London), 1986.

Centre Georges Pompidou, *Jean Prouvé*, Centre Georges Pompidou (Paris), 1990.

Centre Georges Pompidou, *Architecture et industrie*, Centre Georges Pompidou (Paris), 1984.

Daidalos: *Berlin Architectural Journal*, no 18 December 1985.

De Mare, Eric, *The Nautical Style*, Architectural Press (London), 1973.

Deutsches Architectural Museum, *Vision der Moderne*, Prestel (Frankfurt), 1986.

Dickens, AG, *The Age of Humanism and Reform*, Prentice Hall (New York), 1977.

Dini, Massimo, *Renzo Piano*, Architectural Press (London), 1984.

Foster Associates, *Foster Associates' Buildings and Projects*, vol 2, Watermark (London), 1989.

Frishmann, Martin and Hasan Uddin Khan (eds), *The Mosque*, Thames and Hudson (London), 1994.

Gill, Brendan, *Many Masks*, Heinemann (London), 1988.

Hayward Gallery, *Le Corbusier: Architect of the Century*, Hayward Gallery/Arts Council (London), 1987.

KF Schinkel, Collected Architectural Designs (facsimile of Schinkel's collected drawings), Academy/St Martin's Press (London/New York), 1982.

Jencks, Charles, *Post-Modern Neo/Classicism*, Architectural Design (London), 1981.

Lafont, R (ed), *Great Works of Mankind*, Bodley Head (London), 1961.

Los, Sergio, *Carlo Scarpa*, Taschen (London), 1993.

Moore, Rowan, *Nicholas Grimshaw and Partners: Structure, Space and Skin*, Phaidon (London), 1993.

Picou, Antoine (ed), *L'art de l'ingénieur*, Pompidou Centre (Paris), 1997.

Rudofsky, Bernard, *The Prodigious Builders*, Secker and Warburg (London), 1977.

Rykwert, Joseph, *The Dancing Column*, MIT Press (Cambridge, Mass), 1996.

Sandaker and Eggen, *The Structural Basis of Architecture*, Phaidon (London), 1993.

Scarre, Colin, *Timelines of the Ancient World*, Dorling Kindersley (London), 1993.

Sereny, Gitta, *Albert Speer: His Battle with Truth*, Picador (London), 1995.

Siegel, Curt, *Structure and Form in Modern Architecture*, Crosby Lockwood (London), 1962.

Siliotti, Antonio, *Egypt: Temples, Men and Gods*, Thames and Hudson (London), 1994.

Spens, Michael, *Le Grand Bleu Marseille*, Academy Editions (London), 1995.

Stamp, Gavin (ed), 'Britain in the Thirties', *Architectural Design* (London), 1980.

Summerson, John, *The Classical Language of Architecture*, Thames and Hudson (London), 1988.

Tzonis, Alexander and Liane Lefaivre, *Architecture in Europe since 1968*, Thames and Hudson (London), 1992.

Venturi, Robert, *Complexity and Contradiction in Architecture*, MOMA (New York), 1966.

Welsh, Charles, *History of the Monument*, (London), 1893.

Weston, Richard, *Villa Mairea: Alvar Aalto*, Phaidon (London), 1992.

Wood, John Edwin, *Sun, Moon and Standing Stones*, Oxford University Press, 1978.

Zerbst, Rainer, *Gaudí: Arte y Arquitectura*, Rikuyo-sua (Tokyo), 1985.